Introductory Physics of the Atmosphere and Ocean

Introductory Physics
of the
Atmosphere and Ocean

by

L. Hasse
Institut für Meereskunde, Universität Kiel, F.R.G.

and

F. Dobson
Bedford Institute of Oceanography, Dartmouth, Canada

SPRINGER-SCIENCE+BUSINESS MEDIA, B.V.

Library of Congress Cataloging in Publication Data

Hasse, L.
    Introductory physics of the atmosphere and ocean.

    Includes index.
    1. Dynamic meteorology. 2. Fluid dynamics.
3. Oceanography. I. Dobson, F. II. Title.
QC880.H35   1986   551.5   85-24437

ISBN 978-90-277-2139-6   ISBN 978-94-009-5484-7 (eBook)
DOI 10.1007/978-94-009-5484-7

Reprinted with corrections and revision from
"Air-Sea Exchange of Gases and Particles" edited by P. Liss and G. Slinn (NATO 108)
© 1983 D. Reidel Publishing Company

CONTENTS

FOREWORD

The two chapters of this book originally appeared in "Air-Sea Exchange of Gases and Particles", edited by P.S. Liss and W.G.N. Slinn. We wrote them as a general introduction to the physical processes in the atmosphere and ocean which govern the transport of gases and particles in and between the two media. Our audience was to be graduate students in physical chemistry of air and sea, and research workers wishing to get started in this or a related field.

It was Dr. Alan Longhurst, Director-General of the Atlantic Region, Canada Department of Fisheries and Oceans, who pointed out that our introduction had a far wider audience: in fact, anyone with a scientific background who needs a basic understanding of the physics of the atmosphere and ocean. Dr. D.J. Larner of Reidel agreed, and this book is the result.

Since we expended considerable effort to satisfy the demands of the physical chemists, and also discussed the explanations much with our colleagues at home, we expect the reader will find the two parts to be complementary and useful as a unified reference text. On the other hand, since it was designed as background material for a text on air-sea gas exchange and transport, the more experienced reader will be aware that the picture presented emphasizes transport and exchange processes while it ignores others. No mention is made, for example, of weather forecasting; neither is large-scale ocean modelling considered. On the other hand, extra attention is paid to atmospheric transport and to ocean surface processes. The references reflect this same bias, although the separately collected "General References" contain excellent material for anyone wishing to read more deeply in any relevant area.

ACKNOWLEDGEMENTS

The book in which these introductions appeared was originally commissioned by the NATO Science Committee Special Programme Committee on Air-Sea Interaction. The members of that committee made extensive use of their powers of persuasion to convince Peter Liss and George Slinn to co-edit the textbook, and then

worked with Liss and Slinn to persuade the various contributors
(all active and busy people) to present their sections for criticism
at a NATO Advanced Study Institute hosted by the University
of New Hampshire at Durham, N.H. in July, 1982. (Our articles
benefitted greatly from these critiques.) The work of both
the Air-Sea Interaction Committee and the editors was ably assisted
by Dr. Mario Di Lullo of the NATO Secretariat. The present
volume has been helped along by Mr. M. Latremouille of BIO.

We both wish to express our indebtedness to the many
colleagues and friends at the Institut für Meereskunde and the
Bedford Institute of Oceanography who, directly or indirectly,
helped us to explain things as well as we have, and supported
us while we did it.

L. Hasse and F. Dobson   July 26, 1985

INTRODUCTORY METEOROLOGY AND FLUID DYNAMICS

Lutz Hasse

Institut für Meereskunde an der
Universität Kiel
D 2300 Kiel, B.R. Deutschland

1. SOME DEDUCTIONS FROM BASIC PHYSICS

Atmosphere and ocean are probably the most complex system which is treated in physics, since the scales of motions involved span such a wide range:some 10,000 km for long planetary waves, through a few millimeters for the scale of decaying turbulent eddies. The motions at different scales interact. An overview of such a complex system is necessarily incomplete. Emphasis is on such processes which relate more or less directly to the transport of gases and particles. The treatment in the introductory chapters on meteorology and physical oceanography is descriptive rather than deductive.

Meteorologists and oceanographers describe the motion and thermodynamics of the atmosphere and the ocean by use of the basic laws of physics. Hence, the understanding is based not only on observations, but also on physical reasoning and theoretical deductions from first principles. There is conservation of momentum, conservation of mass (for air, water vapor, water and salt), conservation of energy and equations of state for ideal gases and a thermohaline sea.

The balance of momentum yields the equations of motion, also called Navier-Stokes equations. According to Newton's second law; mass times acceleration is equal to the sum of forces acting on the mass. The balance is given in (1) for a volume of fluid, written as an equation for accelerations, that is Newton's law is divided by the mass per volume, the density ρ.

$$\frac{du}{dt} = -\frac{1}{\rho}\frac{\partial p}{\partial x} + fv - \frac{1}{\rho}\frac{\partial \tau_{xz}}{\partial z} \tag{1.1}$$

$$\frac{dv}{dt} = -\frac{1}{\rho}\frac{\partial p}{\partial y} - fu - \frac{1}{\rho}\frac{\partial \tau_{yz}}{\partial z} \tag{1.2}$$

$$\frac{dw}{dt} = -\frac{1}{\rho}\frac{\partial p}{\partial z} - g \tag{1.3}$$

Velocity components u,v,w are given in a cartesian coordinate system with z vertical so that the acceleration due to gravity, g, is in the negative z direction. The other terms on the right hand side are the acceleration due to the pressure gradient, the Coriolis acceleration and the acceleration due to friction. Newton's law is valid only in an unaccelerated coordinate system. Since we are measuring in a coordinate system rotating with the earth, a fictious acceleration appears, called Coriolis acceleration or acceleration due to the rotation of the earth. A motion which is straight and unaccelerated in an inertial (unaccelerated) system will appear as curved to an observer rotating with the earth. Looking with the flow, the deflection is to the right on the Northern Hemisphere and to the left on the Southern. The factor

$$f = 2\Omega\sin\phi, \tag{2}$$

where ϕ is latitude and Ω rotation rate of the earth, is called the Coriolis parameter. Hence the Coriolis acceleration vanishes at the equator where f changes its sign.

The third term on the right hand side is the friction term. τ denotes a stress that is a force per area. The index xz denotes that a force in x direction acts on a plane with the normal in z direction. The difference of the stresses at the sides of any volume is felt as a frictional force by the volume. The stress may also be considered as the x-momentum going through the plane with normal z per unit time and area. This momentum transport is brought about by molecular movements (Newtonian friction). The equations of motion are derived for the forces at a volume in a given moment. They are also used for the 10 minute means of velocity as usually determined in routine measurements. If used

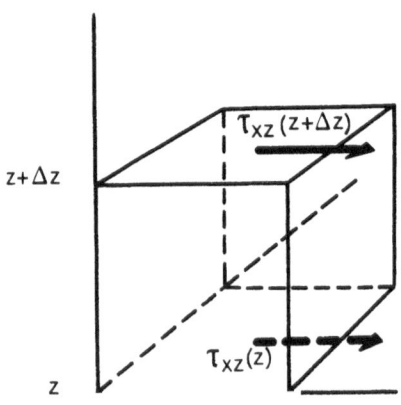

for mean velocity components, the shorter scale motions also transfer momentum and, therefore, additional stresses, the turbulent or
Reynolds stresses, appear. The stress components are then the sum
of turbulent and molecular stress, and so is the friction term.
Except very near to the interface (see the interface section below) the turbulent stresses are much larger than the molecular
and the friction term is essentially due to turbulence.

The left hand side of (1) gives the acceleration of a fluid
parcel. Since the forces operate on the mass of the parcel, this
is called a "substantive" or "material" derivative. This is the
derivative following a moving parcel. Since measurements are more
often done at one point over a time or at one time over a spatial
distance, Euler's relation (3) is used heavily,

$$\frac{dp}{dt} = \frac{\partial p}{\partial t} + u\,\frac{\partial p}{\partial x} + v\,\frac{\partial p}{\partial y} + w\,\frac{\partial p}{\partial z} \qquad (3)$$

where p stands for any meteorological or oceanographical variable
p(x,y,z,t) like a temperature or velocity component.

The notations $\partial p/\partial t$, $\partial p/\partial x$, $\partial p/\partial y$, $\partial p/\partial z$ denote partial derivatives with respect to time and the three spatial coordinates.
The notation implies that although p is a function of four variables, the remaining three are held constant when differentiating.
The partial derivatives are also called local derivatives in contrast to the substantive derivative. The last three terms on the

Scale	Inertial terms horiz.	vert.	Pressure term	Coriolis term	Friction term
OCEAN Large-scale	10^{-8}	10^{-8}	10^{-5}	10^{-5}	10^{-8}
Meso-scale	10^{-4}	10^{-5}	10^{-4}	10^{-4}	10^{-5}
Small-scale	10^{-2}	10^{-2}	10^{-2}	10^{-4}	10^{-2}
ATMOSPHERE Synoptic-scale	10^{-4}	10^{-4}	10^{-3}	10^{-3}	10^{-5}
Meso-scale	$3 \cdot 10^{-4}$	10^{-4}	$3 \cdot 10^{-4}$	$3 \cdot 10^{-4}$	10^{-5}
Cumulus-scale	10^{-3}	10^{-3}	10^{-3}	$3 \cdot 10^{-4}$	10^{-4}
Small-scale	10^{-4}	10^{-4}	10^{-3}	10^{-3}	10^{-3}

Table 1. Magnitude of terms (in m/s^2) in the horizontal
equations of motion for various scales of motion in
the ocean and the atmosphere.

right are called the advective terms, since if there is a gradient

$$\frac{\partial p}{\partial x}, \frac{\partial p}{\partial y}, \frac{\partial p}{\partial z}$$

of the property p, this will be advected (transported) with the flow, given by the velocity vector (u,v,w). The local rate of change at a point is determined by

$$\frac{\partial p}{\partial t} = \frac{dp}{dt} - \left(u \frac{\partial p}{\partial x} + v \frac{\partial p}{\partial x} + w \frac{\partial p}{\partial z} \right) \tag{3}$$

that is, by the rate of change occurring in the parcel which happens to be at the point, minus the rate of change due to the fact that parcels with different values of p are carried over the point by the flow of the fluid.

If (3) is used to rewrite the terms on the left side of (1), it is seen that the Navier-Stokes equations are non-linear, that is, the variables u,v,w appear as products and squares. This indicates that the different scales of motion will interact. Hence, we have a very complex interactive system. With the presence of small scales and the interaction between scales, the atmosphere and ocean possess a certain degree of unpredictability. (Small scale in this sense means smaller than we could ever resolve in experimental determination of an initial state of the atmosphere and ocean or in their numerical modelling. This includes scales as large as cumulus clouds.)

The equations of motion contain the density ρ. In meteorology, it is customary to measure air temperature and pressure, while the density or specific volume is not. The mixture of gases that constitute the dry atmosphere behave like an ideal gas. Hence, the equation of state

$$p/\rho = RT \tag{4}$$

(where R is the individual gas constant of dry air) can be used to calculate density from temperature and pressure. The effects of humidity are taken into account separately. In the case of sea water the density depends, in a weakly nonlinear way, on both its temperature and its concentration of dissolved salts, referred to as its "salinity" (See Dobson, this volume.)

In large-scale flow the vertical accelerations are small in the sense that dw/dt can be neglected against g. Hence, the vertical component of the Navier-Stokes equations reduces to the hydrostatic equation:

$$\frac{\partial p}{\partial z} = - g\rho \tag{5}$$

In static equilibrium the pressure is the weight of the fluid in the vertical column over any area. Hence, the pressure field is given by the density field. The equation of state for air tells us that for a given pressure the density depends only on the temperature. Hence, the wind field is known from the temperature field, except near the surface where a deviation is forced by friction.

The situation is similar in the ocean in the sense that temperature and salinity determine the density field and hence the pressure field. This dual dependence means that flows exist in the ocean which are driven by the influence of both temperature and salinity; they are called "thermohaline" circulations. Yet, determination of flow by aid of the Navier-Stokes equations is more difficult in the ocean than in the atmosphere. The hydrostatic equation is a differential equation and must be integrated with height to obtain a relationship between pressure and height. In the atmosphere the integration is done from the mean sea surface level. In the ocean such reference level is not available and oceanographers must infer the absolute velocity by other means.

The equations of motion, as given in (1), are slightly simplified, but perfectly sufficient for the purpose of this book. An analysis of the order of the terms is given in Table 1. In many cases even simpler forms can be used, see for example the hydrostatic equation (5) as an approximation of the vertical component of the equations of motion. This is a good approximation for all except the smallest scales of motion (that is for most oceanic and atmospheric flows except for convection, surface and internal waves, and turbulence).

For the horizontal components (1.1, 1.2) the simplest type of equilibrium flow is a balance of the pressure gradient and the Coriolis accelerations:

$$u_g = -\frac{1}{\rho f}\frac{\partial p}{\partial y}$$
$$v_g = +\frac{1}{\rho f}\frac{\partial p}{\partial x} \tag{6}$$

This flow is called the geostrophic wind or current. The geostrophic equilibrium is a good approximation for the velocity field in the free atmosphere (i.e. above the influence of surface friction) and in the interior of the ocean, away from boundaries. It provides for the flow to follow the direction of the isobars (lines of constant pressure). Hence, in the atmosphere we have the typical flow direction around lows and highs as indicated in Figure 1. Geostrophic flow is a helpful device, since it relates the wind or current in a simple way to the pressure field.

Warm fluid is less dense than cold fluid. As the Figure 2 shows, a horizontal temperature gradient yields different horizontal pressure gradients at different heights. Consequently the geo-

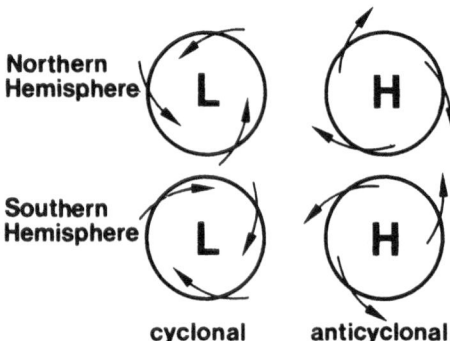

Figure 1. Circulation pattern around highs and lows at the Northern and Southern Hemisphere. In this idealised sketch the circles are isobars. The wind arrows would be parallel to the isobars in the free atmosphere. Convergence into the lows and divergence out of the highs is found at the surface due to friction.

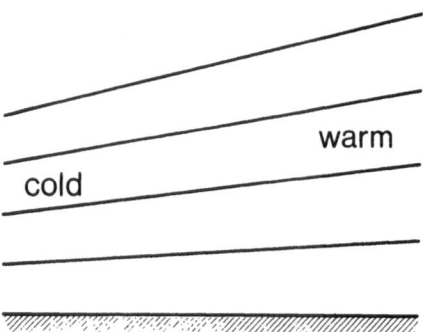

Figure 2. Sketch showing the effect of horizontal temperature gradients on the height of isobaric surfaces (full lines) in a cross section. The increasing slope of the isobaric surfaces with increasing height yields, in geostrophic equilibrium, an increasing wind component in the direction into the paper (in the Northern Hemisphere, reversed direction in the Southern Hemisphere).

strophic velocity varies with height. This change of geostrophic flow with height is called the "thermal wind", since it is related to the horizontal temperature gradient. It is also called the "geostrophic shear". In the atmosphere the most prominent example of the thermal wind is at the strong temperature gradients of frontal zones. The polar and subtropical jet streams are found at the top of these frontal zones. Similar fields are caused in the ocean by salinity and temperature (i.e. density) gradients.

Though in the free atmosphere the geostrophic equilibrium is a good approximation, the small deviations from this equilibrium are responsible for the change of weather. Also, the friction at the surface has considerable influence on weather. Near the surface, there is an approximate equilibrium between Coriolis acceleration, pressure gradient and friction. Friction yields a reduced wind speed compared to geostrophic flow and a deviation of the wind direction from the isobars towards low pressure. Hence, near the surface there is an outflow from a high and an inflow into a low (see Figure 1). This results in ascending motion at low pressure and descending at highs. Much of the motion in the ocean is in geostrophic balance but, like the atmosphere, small deviations from the balance have important effects when superimposed on the geostrophic flow. In the surface waters, where friction is important, the wind drives geostrophic "Ekman" flows which diverge and converge in response to large-scale variations in the strength and direction of the wind; such effects in turn induce the large-scale gyral motions in the ocean.

Conservation of mass is used to derive the continuity equation

$$\frac{\partial u}{\partial x} + \frac{\partial v}{\partial y} + \frac{\partial w}{\partial z} = -\frac{1}{\rho}\frac{d\rho}{dt} \tag{7}$$

This simply states that if the volume occupied by a given mass is changed, its density must change in reverse. The continuity equation is mostly used in the incompressible form, that is for constant density $d\rho/dt = 0$. This is correct in the sense that the effect of density variations is smaller than that of velocity variations. Under this condition, the continuity equation can be used to calculate vertical velocities from horizontal convergences or divergences.

To understand why ascending and descending motions in the atmosphere produce quite different effects we have to mention a few other processes.

Conservation of energy is used to derive the first law of thermodynamics. This says in the simple form

$$dU = dQ - pdV \tag{8}$$

that for a closed thermodynamic system the rate of change of internal energy dU is proportional to the rate at which thermal energy
dQ is applied to the system minus the rate at which mechanical
work is done. The mechanical work is typically the work done by
compression or expansion of fluid under the pressure of the fluid
around it. Since, in meteorology, density or specific volume is
usually not measured, the first law is rewritten in terms of temperature and pressure as

$$\frac{dQ}{T} = c_p \cdot \frac{dT}{T} - (c_p - c_v) \frac{dp}{p} \tag{9}$$

An important application of this equation is for vertical
movements of fluid parcels. Since there are always strong vertical
pressure gradients in the air and sea water, a vertical movement
will lead to a pressure change, which, by virtue of the ideal gas
equation for air and the equation of state for water, will result
in a change of volume and/or temperature. The most simple case is
the one where no thermal energy is supplied, called an "adiabatic
process". In this case in the atmosphere there exists a unique
relationship between pressure and temperature, called Poisson's
equation. A parcel, which moves vertically under adiabatic conditions, has a "dry adiabatic lapse rate" (temperature drop with
height) of about 1 K/100 m in the atmosphere. Since processes -
aside from condensation/evaporation processes - that provide heat
to an atmospheric parcel typically have a time scale of order one
day, while many vertical movements are of shorter duration, the
assumption of adiabatic vertical movement is often a good approximation. The same applies to the deep ocean, except that mixing
rates are much slower, so the equivalent time scale comparison is
months to days instead of days to hours. In the ocean, the adiabatic lapse rate is about 0.15 K/km at a depth of 5 km.

One very important constituent of the atmosphere is water
vapor. The conservation of mass is also used to obtain a balance
equation for the water vapor content. If $\dot{W}$ is the rate of condensation, and q, specific humidity, then

$$\frac{\partial q}{\partial t} + u \frac{\partial q}{\partial x} + v \frac{\partial q}{\partial y} + w \frac{\partial q}{\partial z} = - \dot{W}/\rho \tag{10}$$

A similar equation holds for the salt content in sea water. (The
equation is derived for a parcel, and the velocity components are
momentary components. The change of property q of the parcel by
molecular diffusion across the fluid boundaries of the parcel is
not considered: if important, that must be included in $\dot{W}$.)

Basic thermodynamics are also used to describe the phase
change of water vapor, fluid water and ice in the atmosphere. The
importance of water vapor in the atmosphere stems from the role of

evaporation-condensation-precipitation in the hydrological cycle, from the radiational properties of water vapor and of liquid water (in clouds), and from the energy transfer with the evaporated water. The heat of vaporisation is 2.5 kJ/g, the heat of fusion 0.33 kJ/g (both slightly temperature dependent). In the absence of liquid water, the water vapor follows the equation of an ideal gas with sufficient accuracy. In the presence of liquid water, there is an equilibrium between the gaseous and the liquid phase. This equilibrium is temperature dependent, as described by the saturation water vapor curve (Figure 3). This Figure is the graphical presentation of the Magnus equation which is derived from the Clausius-Clapeyron equation. The temperature dependence of the saturation

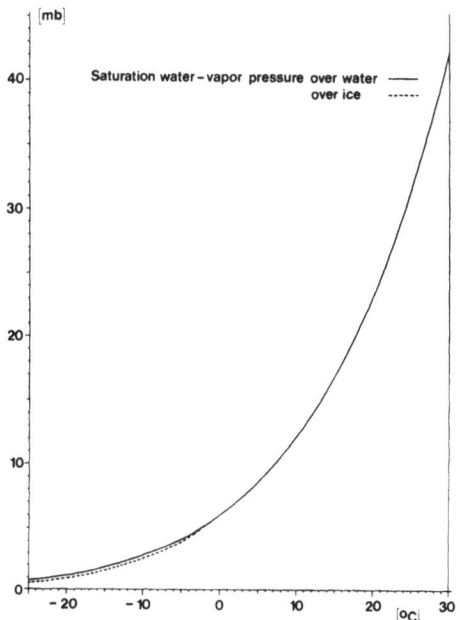

Figure 3. Equilibrium saturation water vapor curve. At vapor pressure below the curve, the relative humidity is less than 100%. At vapor pressure above the curve, the vapor would condense to either liquid water (mist or cloud droplets) or ice crystals (at low temperatures).

water vapor pressure is the reason for the higher water vapor content of tropical air compared to air of the temperate zones and the polar regions. In a similar way, since the temperature decreases on average by about 0.6 K/100 m in the troposphere, the water vapor content is mainly concentrated in the lower 1 km to 3 km. In the ocean there are no phase changes of comparable effect to that of water vapor in the air. Freezing and melting occur in

higher latitudes, but they are highly localized in time and space.
Evaporation at the sea surface increases water density both by
cooling and by increase in salinity, causing convective overturning of the surface waters. Precipitation causes stable stratification by a layer of fresh water on top of the salt water, noticable
at low wind speeds.

The cooling by expansion, together with the water vapor content of the atmosphere, is responsible for most of the "weather".
Consider rising air; by expansion, the temperature of the air is
lowered, but the air takes its moisture with it. With the cooling,
the humidity of the air may reach saturation and condensation will
take place with a release of latent heat. Cloud droplets and eventually rain drops will be formed. This is the typical case of
cumulus convection and of the ascending air in a depression. The
opposite case is found in the free air between cumulus clouds and
in high pressure centres. The sinking air is heated by adiabatic
(or nearly adiabatic) expansion, clouds evaporate and the air becomes dry. There is no analogous process in the ocean.

The role of the Coriolis acceleration in atmospheric and
oceanic flow is quite interesting. Consider a container filled with
fluid of different temperatures (densities) in an unaccelerated
system as in Figure 4a. If the divider is removed, the pressure
difference at the interface would produce a motion and in the final
stage (Figure 4c) the warm fluid would be stratified above the
cold fluid. Some of the system's potential energy has been converted into kinetic energy. The initial situation (without divider)
is unstable.

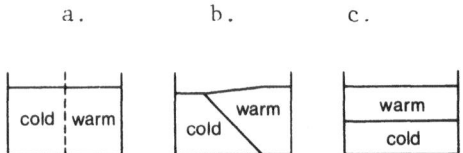

Figure 4. Neighbouring cold and warm fluid is unstably
stratified. In a non-accelerated system, the resulting
pressure gradient would drive the fluid until warm is
above cold (a→b→c). At the rotating earth a sloping
front, like (b), can be stable between moving cold and
warm fluid, since the Coriolis force can balance the
pressure gradient force.

On the rotating earth, in large-scale flow, if we neglect
friction, pressure gradients are balanced by the Coriolis acceleration. Hence, the flow is along the isobars and parallel to the
interface between cold and warm. A situation like Figure 4b can
be in equilibrium (that is, no flow perpendicular to the interface)
with geostrophic flow on both sides parallel to the interface.

This flow must fulfil a boundary condition similar to the thermal wind condition depicted in Figure 2 (on the Northern Hemisphere, the wind component into the paper increases with height). This is remarkable, since in this way potential energy can be accumulated when temperature differences due to differential heating build up.

If the horizontal temperature gradients become too large, the equilibrium becomes unstable. That means, if a small disturbance is superimposed on the geostrophic flow, part of the potential energy is used and the disturbance is amplified, its kinetic energy is increased. The typical depressions of the midlatitudes (see below) develop. This instability is called baroclinic instability. 'Baroclinic' indicates, that the surfaces of equal temperature are inclined against the pressure surfaces. This is the case in the center of Figure 2 and in Figure 4a and b. The situation with temperature and pressure surfaces parallel, as in Figure 4c, is called barotropic. Baroclinic instability is an important feature of the large-scale flow on the rotating earth.

2. LARGE-SCALE ATMOSPHERIC CIRCULATIONS

The original source of energy for all atmospheric motion is the radiation from the sun. Of the solar radiation impinging on the outer edge of the atmosphere, about 30% (the so-called planetary albedo) is returned to space by reflection and scattering in the atmosphere and reflection at the ground. The averaged daily flux of solar energy on a horizontal plane at the top of the atmosphere is given in Figure 5. Due to the solar declination, the annual amount of energy received at the equator is only larger by a factor of 2.4 than that at the pole.

The energy received as solar (or short-wave) energy from the sun by the atmosphere-plus-earth system must be returned to space as terrestrial (or long-wave) radiation (Figure 6). The surface of the earth emits according to Stefan-Boltzmann law, that is, the terrestrial radiation increases with the fourth power of the surface temperature. Since some atmospheric gases (especially water vapor and carbon dioxide) and aerosols (airborne particles and water droplets of clouds) absorb and emit long-wave radiation, the earth's surface also receives energy from the atmosphere, the back-radiation. At the same time, the atmosphere radiates towards space.

In total, the atmosphere emits more energy than it absorbs short- and long-wave radiation. The heat balance is closed by the transfer of sensible heat and latent heat (via evaporation) from the earth's surface to the atmosphere. That is, the solar radiation received at the ground is used partially to heat the air and to evaporate water. The heat used for evaporation is released to the

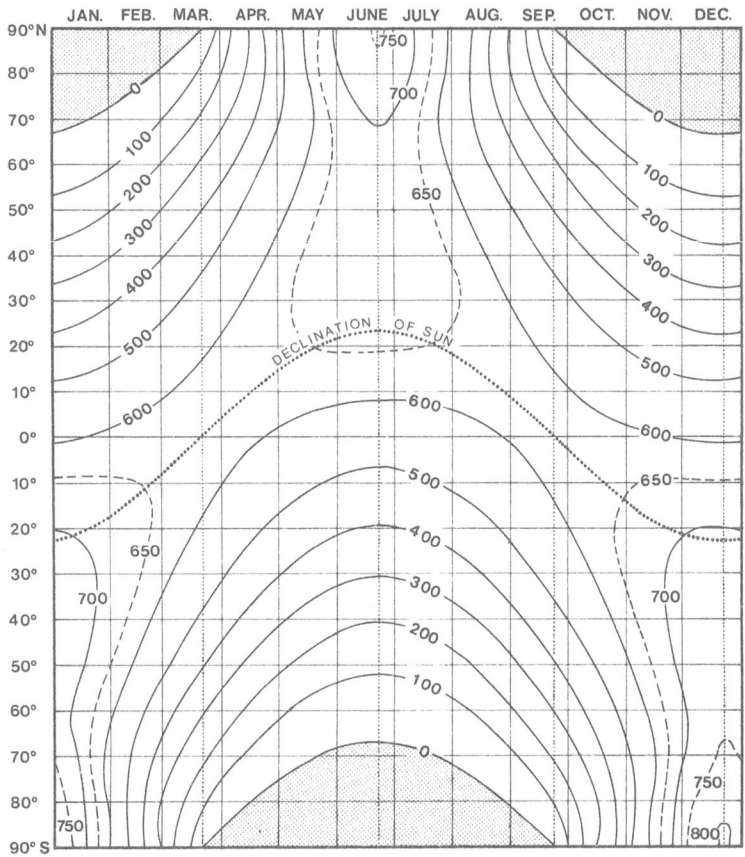

Figure 5. Daily average of solar radiation impinging
on a horizontal surface at the outer edge of the atmosphere, in W/m^2. Solar constant is 1360 W/m^2. Adapted
from List, 1956.)

atmosphere when the water vapor condenses in clouds or fog. It is
thus carried with the air as so-called latent heat. Transport of
sensible and latent heat from the surface by turbulence and order-
ed motions closes the energy balance of the atmosphere, as shown
schematically in Figure 7. Note, however, that such graphs are
meant to show the relative importance of the processes: they give
a global picture, and are not meant to be applicable to any specific place and time. The latitudinal and seasonal variations in
this energy balance lead to corresponding temperature gradients
which drive the atmospheric circulations.

For an understanding of atmospheric circulations it is important to recognize that water and land surfaces have different

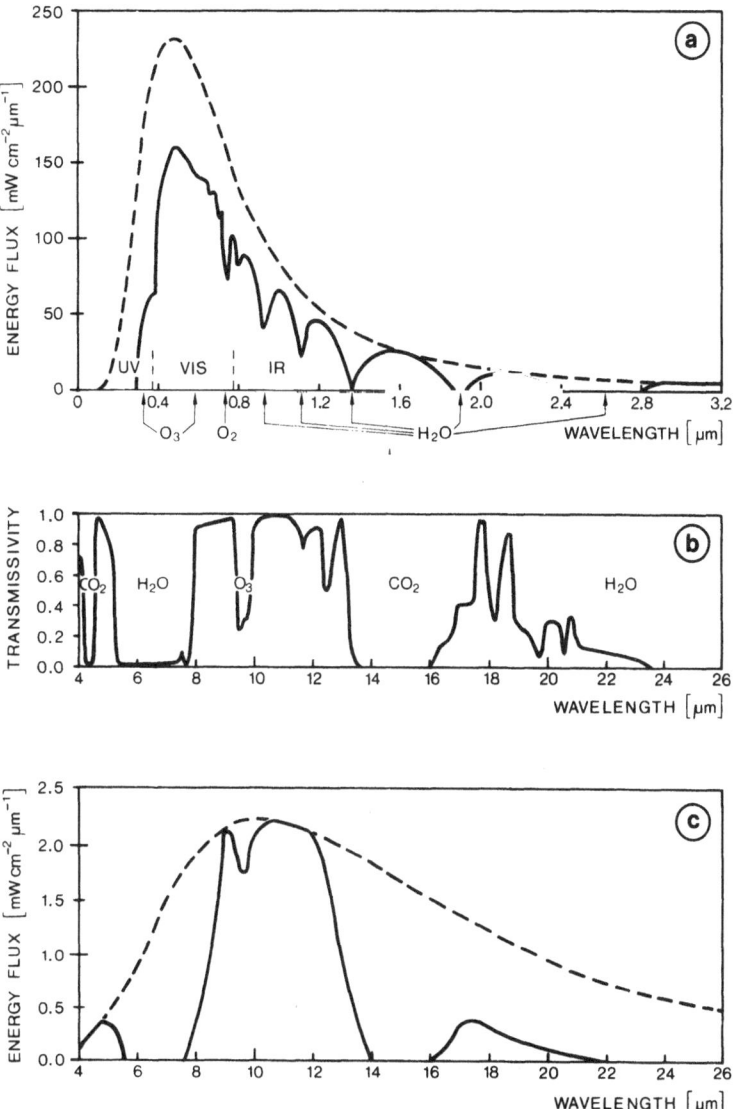

Figure 6. Spectral distribution of solar and terrestrial radiation. The figure is simplified, the very detailed structure of absorption as function of frequency is not considered. (a) Dashed line gives the solar radiation (assumed to be a black body at 6000 K) as received at the outer edge of the atmosphere. Full line is the solar radiation received at the ground after absorption and scattering in the atmosphere. The main absorbers are indicated below the abszissa. (b) Transmissivity of the atmosphere in the infrared. (c) Terrestrial net radiation

at the ground (full line). Net radiation is the balance
between upwelling radiation from the surface minus downwelling radiation from atmospheric gases and particles.
The dashed line gives the black body radiation of the
surface.

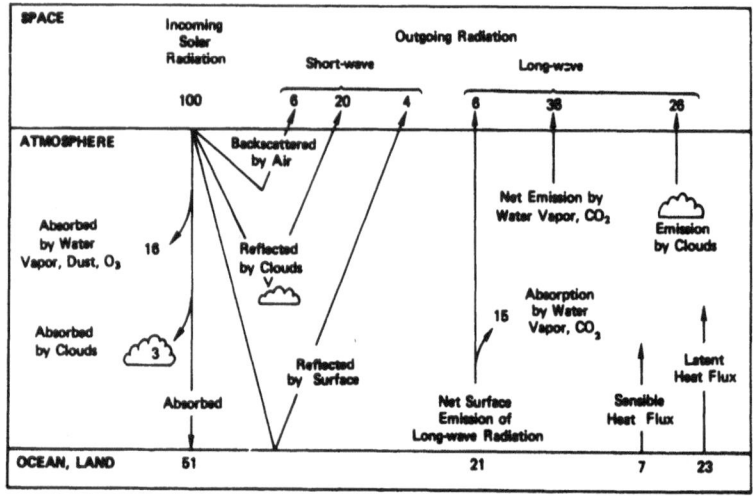

Figure 7.   Energy redistribution in the earth-plus-
atmosphere system. The energies are given as fractions
of the solar radiation impinging on a horizontal surface at the outer edge of the atmosphere. (After NAS,
1975.)

properties. This concerns radiation, heat and moisture storage,
and roughness. The albedo of natural water surfaces, about 6%, is
the lowest of any natural surface (Figure 8). The heat storage is
also quite different for water versus solid surfaces (Tables 2, 3).
In water, part of the solar energy penetrates to several meters
depth. Also, in water turbulent motions distribute the heat absorbed in the uppermost few meters throughout a deeper layer (the
"mixed layer"). For land, the main way to transport absorbed solar
energy is by ineffective molecular conduction. Hence, the land surface temperature responds very fast to any changes in the net surface energy balance, while the sea surface responds only very slowly. The third property, which is different over land and sea,
is the friction (Table 4). The sea surface - even though waves
may look fairly rough - is comparatively smooth. The land is much
rougher for most types of surfaces.

So far, we have mentioned only land and sea. Ice and snow are
somewhat different. Snow has the highest albedo of any surface.
Also, for snow, an energy gain is first used to thaw the snow be-

fore the temperature may rise above zero degrees Celsius. The roughness of level snow (and also of sand, e.g. of tidal flats) is nearly the same as of the sea surface. About 25% of the land and water
surface are covered with snow and ice (where again the ice may be
covered with snow).

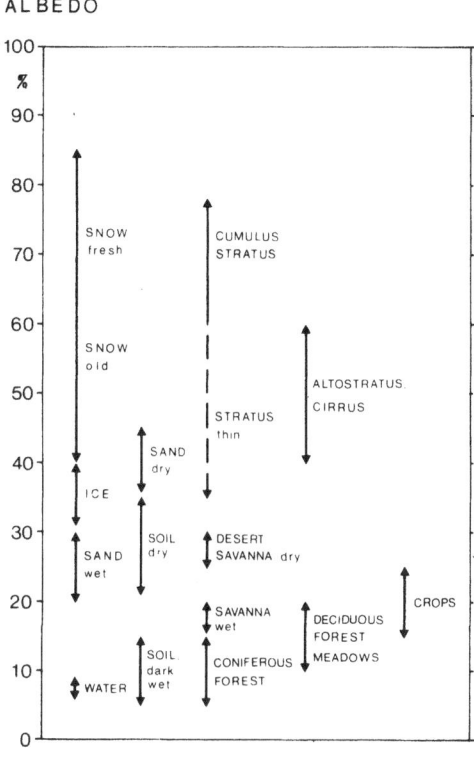

Figure 8. Albedo of different natural surfaces. Albedo
is the shortwave energy returned to space by reflection
and scattering, given as a fraction of incoming solar
radiation.

	Solar radiation albedo	penetration	IR radiation emission coefficient
Water	.06	∿ 50 cm	.92 - .96
Natural surfaces	.10-.30	∿ 1 mm	.90
Snow	.85-.40	∿ 1 cm	.99 - .80
Ice	.30-.40	∿ 10 cm	.96

Table 2. Radiational properties

	ρc $Ws\ cm^{-3}K^{-1}$	κ or K cm^2s^{-1}	$\rho c\sqrt{\kappa}$ or $\rho c\sqrt{K}$
Snow, new	0.13	0.006	0.01
old	0.9	0.003	0.05
Ice	1.9	0.012	0.21
Sand, dry	1.25	0.0013	0.05
wet	1.7	0.01	0.17
Soil	~ 2.6	~ 0.04	0.5
Ocean, very stable	4.2	0.1	1.3
moderately stable	4.2	50	30
well mixed	4.2	300	70
Atmosphere, very stable	$1.2\cdot10^{-3}$	10^3	0.04
near neutral	$1.2\cdot10^{-3}$	10^5	0.4
very unstable	$1.2\cdot10^{-3}$	10^7	4

Table 3. Physical properties of different media. The second column contains either molecular κ or eddy K conductivity as applicable. The third column gives the 'conductive capacity', which takes into account that not only the heat capacity ρc is different for different media, but also that the penetration of a temperature change with depth scales with the square root of conductivity. The larger the conductive capacity, the more energy is needed to produce a given temperature change. (After Priestley, 1959).

	z_0 cm	C_D
sea surface, moderate winds	.015	.0013
high winds	.13	.0020
snow over open natural surfaces	.1	.0019
short grass	1	.0034
long grass	5	.0057
natural surface, open vegetation	10	.0075
forest	20	.010

Table 4. Typical roughness parameters of various surfaces. For definition of the roughness height z_0 and the drag coefficient C_D see section on interface dynamics. The values are given for illustration rather than application. For vegetation there is a tendency to decreasing values with increasing wind. The roughness of the sea surface does not depend directly on wind speed, but rather on sea state. C_D is for 10 m reference height.

The large-scale circulations in the lower atmosphere, the
Troposphere, can be classified into three main regimes (Figure 9):
i) The Hadley Cell, a meridional circulation like a vertically
standing wheel, is driven by the heating at the sea surface. In
its lower branch, the trade winds move over the sea surface, where
part of the sun's energy is used for evaporation and water vapor
is carried towards the equator. The trade winds from the Southern
and Northern Hemisphere converge near the equator in the Inter-
Tropical Convergence Zone, ITCZ. The convergent flow leads to
large cumulus and cumulonimbus clouds. With the rise of the air
the water vapor condenses and releases heat. Hence in the ITCZ
the latent heat collected in the trade wind belt is released and
drives the Hadley circulation. The descending branch of this circulation is found in the subtropical high pressure areas. While
the trade winds are quasi-permanent winds with high directional
stability, the upper poleward branch is less well evident in the
observations (but must have the same mass flow).

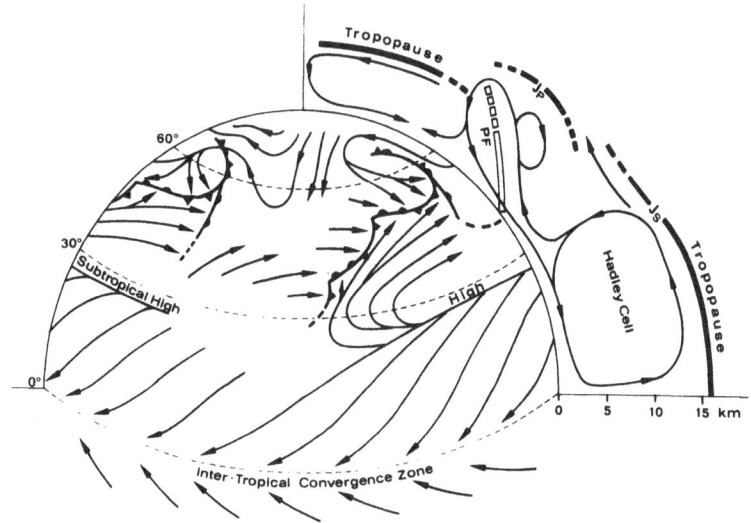

Figure 9. Scheme of the general circulation of the
atmosphere at the surface and in a meridional cross
section. Jp and Js indicate the average positions of
the Polar Front and Subtropical Jet Streams, PF is the
Polar Front. (Adapted from Defant & Defant, 1958.)

ii) The second main regime is the west wind belt of the temperate
zones. The latitudinal gradient in heating leads to temperature
differences at the surface and in the troposphere. The strongest
temperature gradients of the lower troposphere occur in the temperate zones. Such temperature contrasts are found locally over
small distances: the so-called fronts. In principle, from meridio-

nal temperature and, hence, density and pressure gradients one would
expect, in geostrophic equilibrium, a zonal flow. A pure westerly
flow, with meridional temperature gradients exceeding a certain
limit becomes unstable (baroclinic instability). A small disturbance of the flow will be amplified and the zonal flow becomes
wavy. In the middle layers of the troposphere, e.g. at the 500 mb
level, troughs and ridges of low and high pressure are formed
(Figure 10), and in the lower layers the typical disturbances of

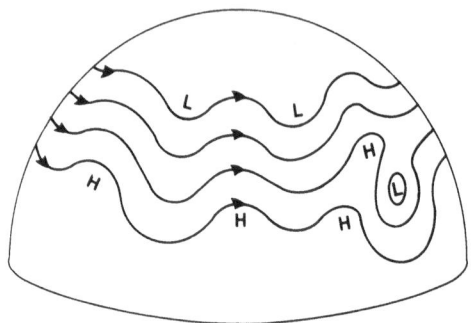

Figure 10.   Diagram of flow patterns at the 500 mb
level (about 5.5 km height).

the temperate zones develop. The result is, that the warm air, originally bordering the cold air, is lifted above it. Potential energy is thus converted into mechanical energy of storms (depressions),
and at the same time, latent heat is released by condensation of
water vapor in the clouds at the fronts. Thus, the two regimes are
quite different, both in their way of converting energy and in
their large-scale transport properties.

iii) A third regime is usually identified: the polar regions are
covered (mostly in winter time) with the so-called Polar High. Since
there is descending air in a high, which must be replenished in the
upper levels, this forms a third, but considerably less energetic
circulation cell.

So far we have emphasized the circulations in the lower atmosphere, the Troposphere. This name denotes the part of the atmosphere which is turned over by weather. Above the Troposphere is
the Stratosphere, which is more stably stratified. But the lower
part of the Stratosphere passively takes part in weather: it is
known from ozone observations that stratospheric air enters the
Troposphere. The Stratosphere shows quite different flow patterns
from the Troposphere: the winter pole has low pressure with westerly winds, and the summer pole has high pressure with easterly
winds above 20 km height. Wind directions reverse in spring and

fall, often rather abruptly.

A summary of the three regimes is given in Figure 11, which gives a meridional cross section of the tropospheric and lower stratospheric circulation. The area of the tropical meridional Hadley cell is seen to have easterly winds at the equator at all heights. The trade winds have an easterly component, while at higher levels westerly winds are found, which at some times or at some places circle the globe with rather high speeds. These are known as the Subtropical Jet Streams. A similar jet stream is found meandering (and with varying intensity) at the polar frontal zone, which divides cold polar air from the warmer subtropical air. Winds are westerly in the midlatitudes, at least on the average. Near the ground, the disturbances may also produce easterly components. At the pole the Polar High in the lower troposphere produces easterly winds, while above a few kilometers the wind is westerly.

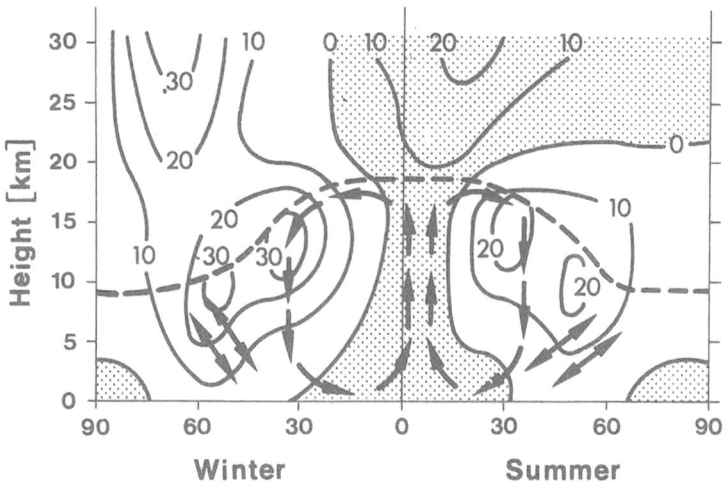

Figure 11. Meridional cross section of the atmosphere showing schematically the zonal winds. Wind speeds are indicated in m/s, blank areas are west winds, shaded areas east winds. The arrows indicate the meridional Hadley cell in the tropics, and double arrows represent the sloping motions of the midlatitude disturbances.

If we look at such a zonally averaged picture, the subtropical highs and the Intertropical Convergence Zone (ITCZ) seem to confine the air to their circulation cells, thus hindering interhemispheric transports and even exchanges between tropical and midlatitudes. This is not true. The zonally averaged view does not allow for the uneven distribution of land and sea over the globe.

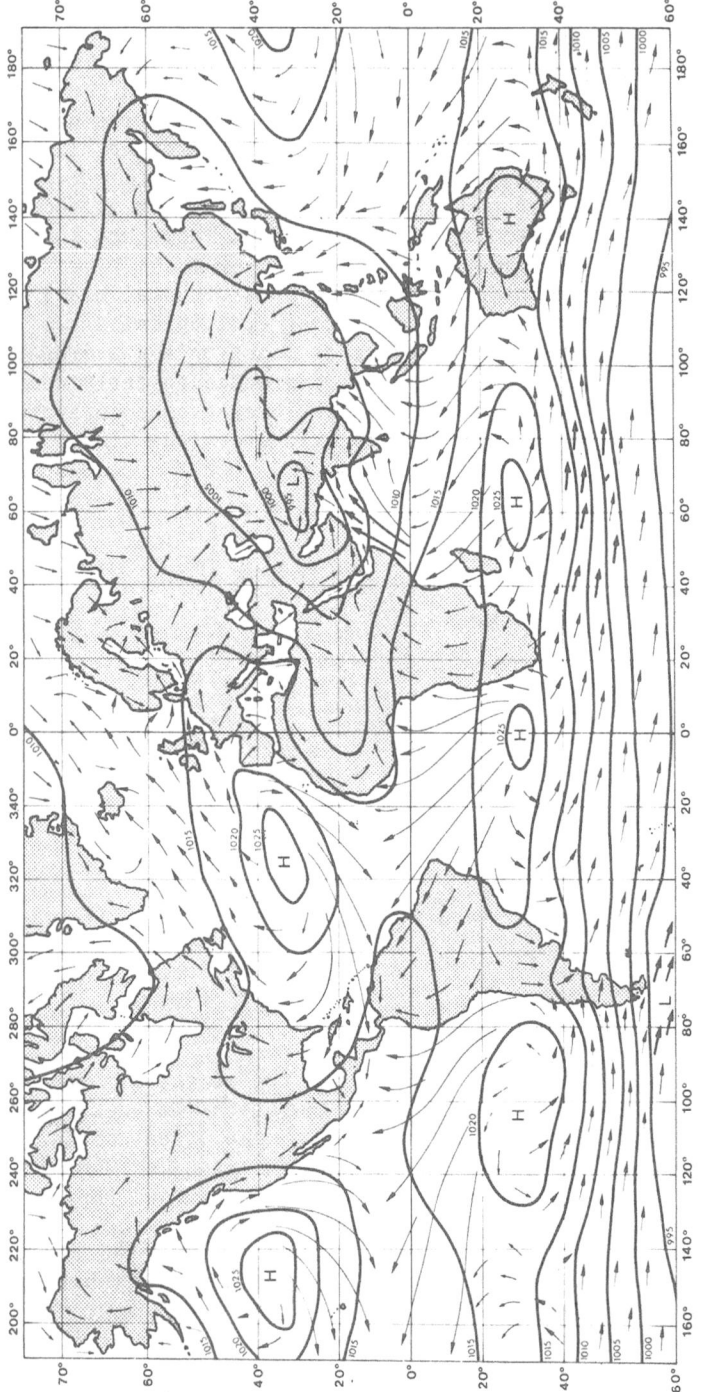

Figure 12a. Mean pressure and winds at the surface in July. Arrows fly with the wind, longer arrows indicate persistent winds. Wind speed code: → 0 to 6 m/s, ⟶ 6 to 12 m/s, ➤ >12 m/s. (Adapted from Köppen, 1899.)

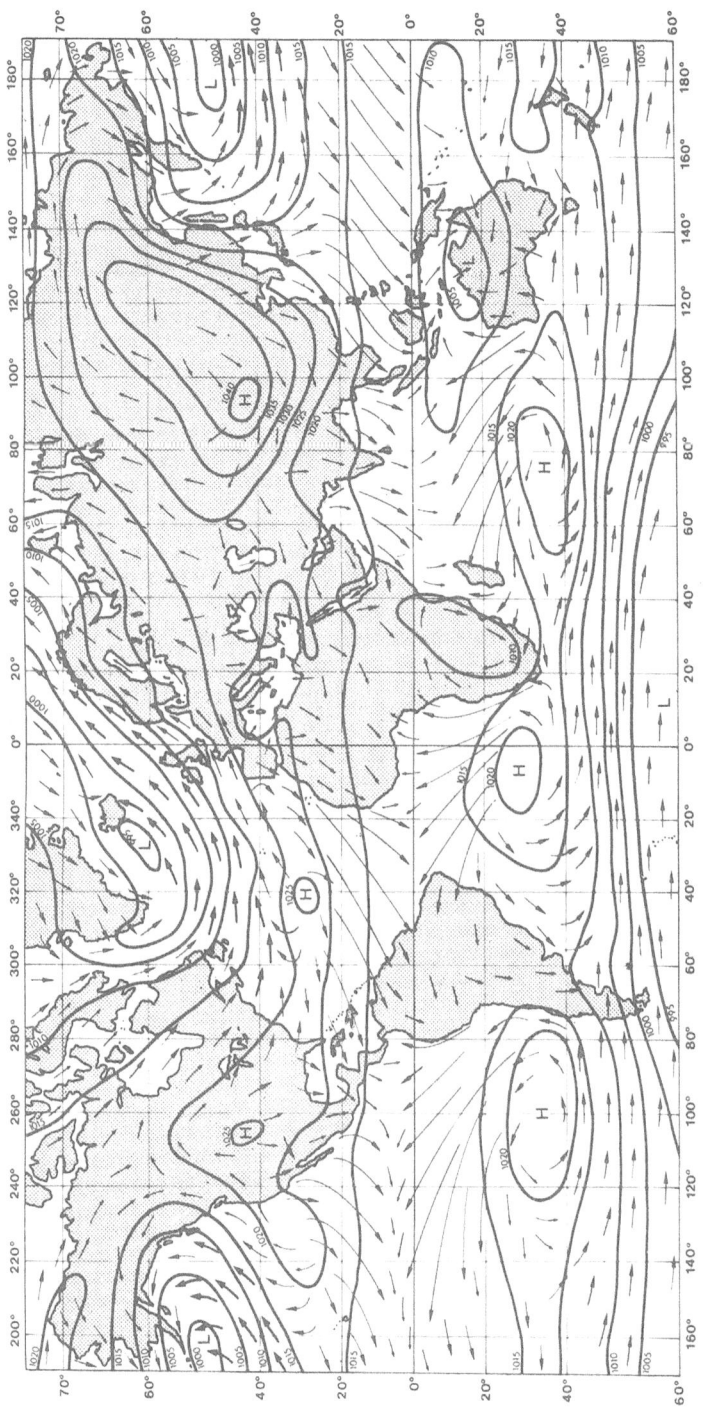

Figure 12b, as Figure 12a, but for January.

Since land and sea have different thermal, radiational, and frictional properties, their distribution modifies the air flow pattern considerably. Consider the wind and pressure field at the surface in summer and winter (Figure 12): In north-summer (July) and south-summer (January) we find high pressure over the oceans and low pressure over the continents. On the Northern Hemisphere the highs are circled clockwise by the wind, on the Southern Hemisphere, counterclockwise (and vice versa for the lows). Low pressure persists at the equator, and the southern midlatitudes have persistently strong westerly winds. Due to the vast masses of the Asian continent, the picture of the Northern Hemisphere is more complicated: there is the winter high pressure center and the summer monsoon low over Asia.

Let me describe some monsoonal effects. The Indian monsoon and similar circulations are brought about by the strong summertime heating. In a more general meaning, the term 'monsoon' is used for any wind system which seasonally reverses wind direction. The monsoon inflow into the Indian sub-continent in summer completely cancels the expected northeasterly trade wind. A similar monsoonal inflow is found into all Southeast Asia. It is also interesting to see that in north-summer the ITCZ is found considerably north of the equator at the Atlantic, while in the area of the Indian Ocean the south-east trade winds turn to a south-west flow and join the monsoonal inflow into India without any ITCZ. In winter the northeast trade winds of the North Atlantic cross the equator and continue as a monsoonal inflow into South America.

It is evident that the transport properties of the lowest 2 or 3 km are not adequately described by a zonally averaged picture. Similarly, in the middle and higher troposphere, the midlatitude westerlies show meandering and considerable deviation from a smooth, confined flow; therefore there is substantial meridional (north-south) transport (Newell et al., 1972). An interhemispheric flow in the middle and higher troposphere and in the lower stratosphere is less evident from the observations, but there is evidence from atmospheric tracers that interhemispheric transport exists.

With respect to atmospheric transport, it is also interesting to look at the global water balance. In units of 1000 cubic kilometers of water per annum (approximately 2 mm precipitation per annum), evaporation at the ocean amounts to 448 and precipitation to 411. The difference (37 units, which is of order 10% of the total evaporation) is carried over land. Over land 65 units are evaporated, and 102 are precipitated. With a water content of the atmosphere of 12400 km^3 and a total precipitation rate of 513 units, the average residence time of water vapor in the atmosphere is about nine days.

3. SYNOPTIC SCALE CIRCULATIONS

The trade winds of the tropical seas are the most persistent wind systems of the atmosphere. Yet there are variations of the wind speed with variations of the positions of the Subtropical High and of the ITCZ. The ITCZ also shows variations of intensity. Cloud clusters travel westward, with clear areas following, so that heavy precipitation is followed by suppressed convection with a period of three days or so.

The disturbances of the tropics are usually classed according to their wind strength as "easterly waves", "tropical depressions", "tropical storms", and "tropical cyclones" (which locally have other names as hurricane, typhoon, etc.). The easterly waves are difficult to identify in the pressure field. A slight trough extends poleward in the isobars of the trade winds. There is convergence before and divergence behind an easterly wave (Figure 13). This produces heavy rain in showers and thunder showers. Tropical depressions are found frequently near the ITCZ

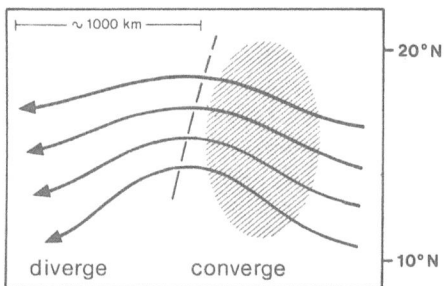

Figure 13.   Diagram of the flow pattern of an Easterly Wave. The convergent flow east of the trough line (dashed) leads to precipitation (shaded area).

and less frequently in the trade-wind belt. Some of them develop to storm speeds, especially in the warm humid season. The tropical depressions have a low pressure center, but still are mostly poorly developed. They are important since they produce heavy rain. The weather in the monsoon areas of India and Southeast Asia is modulated by these depressions. They are also found in the West Indies and drift into Mexico or the Atlantic coast of North America.

Some of the tropical storms develop into tropical cyclones (hurricanes). The tropical cyclones are almost perfectly circular. They start out as unsuspicious easterly waves, squall lines or tropical depressions. In the growing stage, the radius of the

band of hurricane strength winds (say above 35 m/s, i.e.Bft 12
and higher) is only of order 50 km, while in mature stage, storm
strength may extend to 300 km radius. A cross section is given in
Figure 14. Tropical cyclones develop only over warm water and
more than 5 degrees latitude off the equator. Their energy is
taken from the release of latent heat with strong convection.

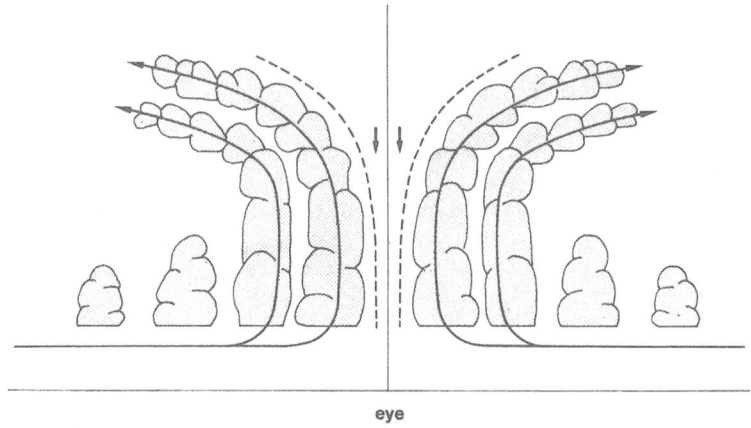

eye

Figure 14.   Schematic cross section of a tropical
cyclone. Hurricane wind strength is found within a
radius of about 80 km, except in the 'eye'. The eye
is cloudless. The storm area has a radius of 300 km
and is filled with convective clouds in spiraling
bands.

Thereby, the storm lives off energy collected in a larger area than
is covered by itself. In the storm area, evaporation is increased
and spray is formed by heavy waves and swell. Tropical cyclones
are seen in satellite pictures as characteristic spiraling cloud
bands and with a cloudless "eye" in the center.

    Weather in the midlatitudes is mainly brought about by travel-
ling depressions. This can be traced back to the latitudinal variation of heating by the sun and the corresponding temperature distribution. The resulting temperature gradients outside the
tropics are often found as strong, localized temperature contrasts,
named fronts. Near the ground, the main temperature contrasts are
found in midlatitudes. This front is taken to divide polar and
subtropical air and is called the Polar Front. Because of the
earth's rotation, neighbouring cold and warm air masses can be in
equilibrium, if certain conditions for the relative velocity components are fulfilled (similar to the thermal wind). Hence potential energy can be built up, which is afterwards released as
energy of storms.

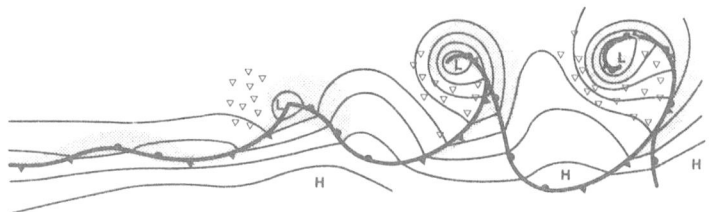

Figure 15. Northern Hemisphere cyclone family according
to Bjerkness and the Norwegian school. Cyclones often
follow each other as disturbances of the Polar Front.
The spatial sequence is also a time sequence with the
youngest developing stage at the left, and the decaying
stage at the right.

The Polar Front can become unstable and disturbances develop.
The life cycle of such a disturbance is depicted in Figure 15. A
cross section of a well developed depression is shown in Figure
16. In the beginning, the cold air was bordering the warm air. In
the developing stage, the warm air glides above the cold air at
the sloping front. This gives rise to condensation and (in the
idealized case) some hours of persistent rain. At the rear, the
cold air advances against the warm air. The scheme shows the cold
air nicely fitted below the warm air. This would mean a fairly
stable situation. The amount of rain would depend only on the velocity with which the warm air is lifted. More often the cold air
advances more rapidly aloft (due to less friction). If cold air
arrives above warm air, the situation is unstable and showers and
thunder showers are formed. The development of a depression stops
when the cold air from the back has reached the cold air in front.
The depression is then filled with cold air near the bottom and the
warm air is lifted above the cold air. Part of the available potential energy and also some of the heat released by condensation
have been converted into the kinetic energy of the depression.
When the energy is spent, the depression is weakened, and the
fronts become inactive and disappear. In the cold air at the rear
of a developed depression, the isobars often converge somewhat
and a line of showers is found. Note that the pictures are exaggerated: the typical slope of a front is 1 in 100. Warm fronts are
inversions in the sense that the normal vertical temperature decrease of 0.6 K/100 m is replaced by a temperature increase with
height.

The high pressure cells provide quite a different weather
type. Friction at the ground forces flow into the low and out from
the high and, by continuity, ascending and descending motions. Ascending motion is destabilizing; descending motion is stabilizing.
The latter provides not only for suppresion of convection in

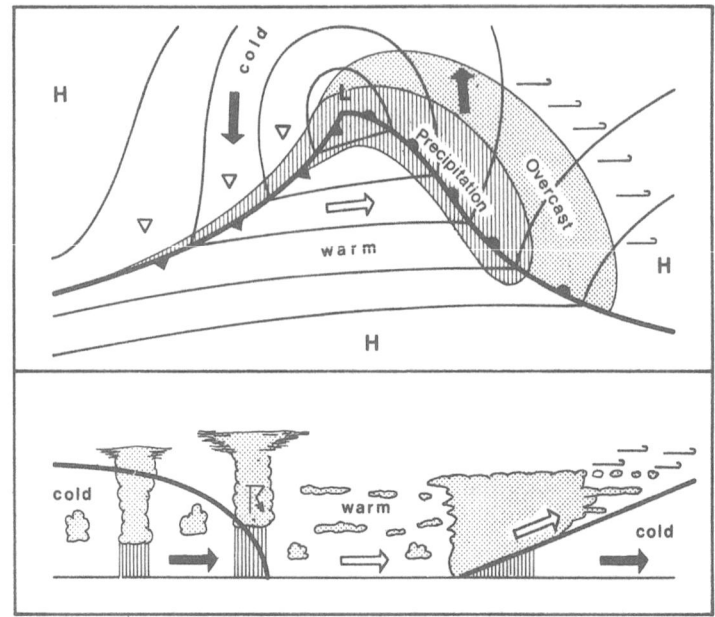

Figure 16.   Cross section (below) and plan view (above)
of a well developed midlatitude depression (Northern
Hemisphere) according to Bjerknes and the Norwegian
school. Triangles indicate showers.

highs, but also for formation of inversions as a typical feature
in the extended high pressure cells. In general, the movements,
development and decay of highs are much slower than those of depressions (except for the intermediate highs in a series of lows). The
highs are rather passive, whereas there is conversion from potential to kinetic energy in the lows.

From a genetic point of view, dynamic and thermal highs are
usually distinguished. Cold air is heavy, and so one would expect
high pressure at the ground in cold air. This is true of the continental (e.g. Eurasian) high in wintertime and the polar high. In
this case the cold air is at the surface (up to 2 or 3 km) only,
and a low is above. The outflow at the ground produces a sinking
motion aloft. The sinking air is warmed by compression and cooled
by radiation, but radiational cooling is more effective at the
ground. The sinking air in the polar highs is dry and naturally
outstandingly clean (both low numbers of nuclei and low humidity,
that is, dry dust only). Recently, however, pollution due to advection of sulfates has increased. On the other hand, high pressure
areas may exist for dynamic reasons, as part of a circulation system. These are warm at the ground due to adiabatic heating from

the sinking motion and high insolation under cloudless conditions.
This is the case for the subtropical highs, which form the poleward,
sinking branch of the Hadley circulation.

4. MESOSCALE CIRCULATIONS

Proceeding to smaller scales of atmospheric motion, we now
come to those circulation systems, which are sometimes called thermal or secondary circulations, or local wind systems. These terms
are used to indicate that temperature differences are responsible
for the onset and existence of local circulations, which are superimposed on the synoptic scale mean flow. There are two well-known
representatives of these local wind systems: the land and sea
breeze systems and the mountain-slope and valley winds. Both are
found mainly with high pressure systems during summer, when the
synoptic scale wind is low and insolation is strong.

The development of a sea breeze due to the temperature difference between land and sea is shown in Figure 17. The reasons
for the different heating of land and sea have been discussed already. The daytime sea-breeze is usually stronger than the nighttime land breeze. The height of the lower branch of the circulation is up to 1.5 km, the height of the returning branch is up to
3.5 km. Typical speeds are 4 to 7 m/s. In the tropics the sea
breeze may penetrate 100 or 200 km inland. These local circula-

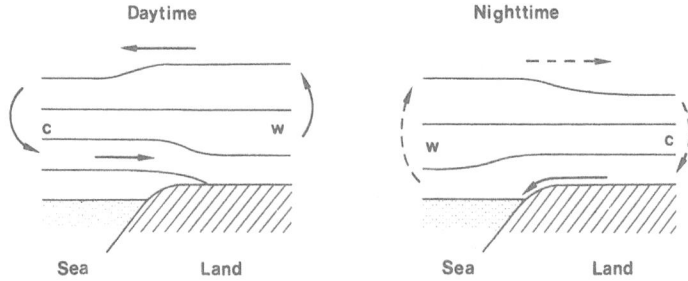

Figure 17. Scheme of the land-sea breeze system. C and
W stand for cooler and warmer. The thin lines indicate
the height of pressure surfaces (vertically exaggerated).
During daytime, when the land is strongly heated by
insolation, the pressure gradients produce an acceleration
in the lower layers towards land and in the higher layers
towards the sea. At night, the land surface is cooled
by outgoing radiation and the pressure difference and
flow direction are reversed.

tions are also found over larger lakes. They are a distinct feature of the weather in the lower latitudes. Here the rising branch
of the secondary circulations may enhance convection and lead to
strong precipitation. A full understanding of such local circulations would also include consideration of roughness differences
and effects of terrain irregularities.

Slope winds are caused by a similar effect. Daytime heating
of a mountain slope produces uphill motion, while nighttime cooling produces downhill flow of cold air. In a similar way the
mean circulation in a mountain valley is up-valley during daytime,
while the cold air flows down the valley during the night (Figure
18). The latter are called the drainage or katabatic winds. They
are often produced on a larger scale by cold air flowing (in winter time) from high lands towards the sea, following the valley
structure. These winds in certain areas attain high speeds and
strong gustiness. They are known by local names (Borha, Mistral,
Santa Ana). Katabatic winds are the predominant surface winds of
the Antarctic continent. Because of the cold surface, the stratification is extremely stable. The cold, katabatic flow has its
maximum already at about 50 m height and may be stronger than the
synoptic-scale wind aloft. A similar wind field is found at Green-
land. Mountain-slope and valley winds are in principle easily ex-
plained. In reality, things seem to be somewhat more difficult.
The gain of solar energy by a slope is very much dependent on its
steepness and orientation and on solar height and azimuth.

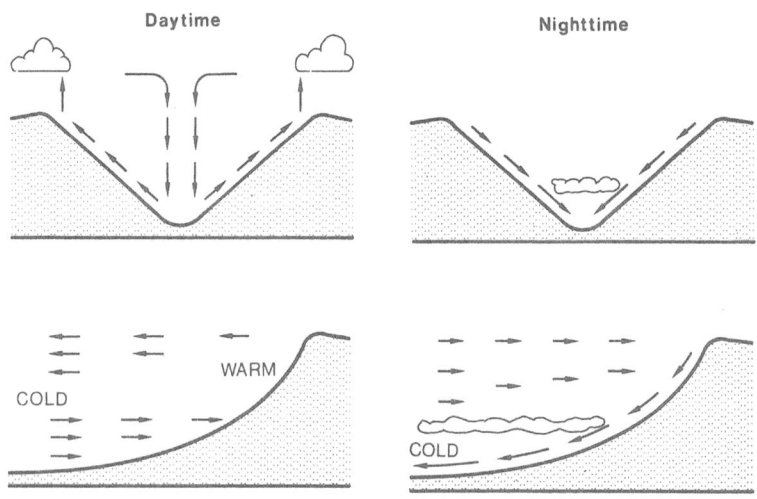

Figure 18. Schematic section of mountain slope and
valley winds.

Also, the local and mesoscale wind field interacts with the synoptic field.

There appear to be very few genuine mesoscale atmospheric circulations on the open ocean(except convection, see below). Coupling between atmosphere and ocean is weak due to the different time and velocity scales, and therefore mesoscale features of the sea surface do not significantly influence processes in the atmophere.

The land-sea transition can influence mesoscale flow in a variety of ways (in addition to the land-sea breeze systems, mentioned above, and perhaps coastal mountain effects). The so-called coastal convergence is produced by the different roughness of land and sea: with an onshore wind, the smaller roughness of the sea permits higher winds than over land. Hence, on crossing the shore line, the flow is retarded. By continuity, a convergence some 10 km inland is found, leading to enhanced convection at some distance from the shore inland and less convection offshore. This is presumably so even over flat terrain, but it is also true that small orographic features (of 50 m height, say) may induce or reduce convective activity through induced upward or downward motions.

The different thermal properties of land and sea produce a horizontal temperature gradient. Even if this does not result in a land-sea breeze system, a thermal wind is imposed that influences the wind profile in the Planetary Boundary Layer. Since the temperature gradient is more or less perpendicular to the shore, the thermal wind will be parallel to the shore. The relationship between the surface wind vector and the large-scale geostrophic wind is accordingly modified. There is evidence that this effect is felt some 30 km offshore.

Turbulence and hence turbulent vertical transports in the Planetary Boundary Layer are also influenced by the advection of turbulence. Consider flow from sea to land; with higher roughness on land, turbulence will be increased. An internal boundary will develop, below which higher turbulence intensity is established, while aloft lesser turbulence levels characteristic of the sea still exist. With increased production of turbulence, the internal boundary will rise within some 10 km to its equilibrium height. In the opposite case (flow from land over the comparatively smooth sea) the advected turbulent energy is larger than in the equilibrium case. The excess energy will mainly be used up by dissipation, which is a relatively slow process. It is estimated that the higher turbulence level is felt even 50 km offshore. This would produce higher diffusion, a larger cross isobar angle and a smaller surface-to-geostrophic wind speed ratio at sea with offshore winds.

Additional effects at a somewhat smaller scale are expected

from bluff shore lines. There is a shadowing effect in the wind
field caused by bluff shores. Channeling effects are also known:
in a valley or an ocean sound between mountains, the flow is parallel to the sound even when the wind aloft and the geostrophic
wind are across the sound. Also, where the sound narrows, higher
wind speeds are attained, which are sometimes in excess of the
geostrophic speed.

Another mesoscale flow is the island or lake effect. Again,
because of different thermal properties of land and sea, an island in the ocean or an inland lake may cause a local circulation.
This is mainly felt when the island or lake is warmer than the
surroundings and induces convection. This may lead to enhanced
precipitation downstream. Also, satellite pictures have given
spectacular evidence that mountainous islands or capes produce
eddies of von Karman vortex street type in the air and water
downstream of the disturbances.

5. CONVECTION AND STABILITY

We now consider the motions of air volumes of smaller extent,
which we may call parcels or blobs, and which may have typical
diameters from say 100 m to a few kilometers. We have described
the larger scale circulations as coherent flows of air masses.
For example at a warm front of a depression the warm air is gli-
ding slowly upwards above the cold air. On the much smaller scale
which we now consider, we must introduce the notion that a parcel
has individual motions, where its characteristic variables may
change. We need to distinguish these from the variables of the
surrounding air, which we conceive as being at rest. The latter
may have a vertical variation of temperature and humidity. These
are called geometric derivatives in order to distinguish them
from the so-called individual changes that a parcel will experience during vertical movements. The motions of a parcel are strongly influenced by the stratification of the ambient air.

Consider a parcel of air. If it is warmer, it is lighter
than its surroundings and will rise, if it is cooler, it is denser
than the surroundings and will be accelerated downwards. Vertical
movements of isolated parcels are accompanied by compression
(downward) or expansion (upward) and an "adiabatic" heating or
cooling of 1 K/100 m results. Hence it is difficult to compare
temperatures from different levels. In meteorology, therefore, a
hypothetical temperature is introduced, which remains constant
with adiabatic movements: the so-called potential temperature.
This is the temperature a parcel of air would have if it were
brought adiabatically to the pressure level of 1000 mb, the approximate pressure at sea level (1 mb = 100 Pa).

Suppose, the vertical temperature profile of the air is measured e.g. by radiosondes. The stratification is stable, when the
potential temperature increases with height (this is the typical
case in the free atmosphere). A parcel originally in equilibrium
with its surroundings will keep its potential temperature when removed adiabatically from its place. At a higher level, it will be
cooler than the surroundings (and at a lower level, warmer)and
hence it will be accelerated to return to its equilibrium level
(Figure 19). In the opposite case, with the potential temperature
of the surroundings decreasing with height, any small adiabatic
move of a parcel from its equilibrium position would render its
temperature such that it would be accelerated away from its original position. This state is called unstable, the vertical stratification is called lapse (the actual temperature decrease with
height is stronger than 1 K/100 m).

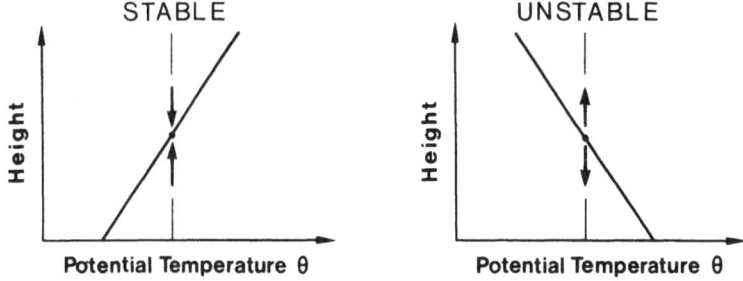

Figure 19. Stable and unstable stratification. The full
line gives the mean temperature profile of the surrounding air. A parcel of air which is originally at the
same temperature as the ambient air keeps its potential
temperature when displaced vertically and hence will be
accelerated either back to its starting point (inversion
condition, stable stratification, left part of the figure)
or will be accelerated away from its equilibrium position
(lapse condition, unstable stratification, right part
of the figure).

An unstable situation is unlikely to persist in the free atmosphere, except near the surface, where the vertical movements (due
to continuity) are hindered by the surface. Here, unstable stratification may persist in a kind of dynamic equilibrium: the stratification is adjusted such that the vertical mixing from the overturning by thermal instability and by mechanically generated turbulence transports heat from the surface at the same rate as it is
supplied. With strong insolation during daytime and infrared radia-
tive cooling during nighttime, there is a strong diurnal cycle of
the net radiation at the surface. The imposed diurnal cycle of surface

temperature produces a typical diurnal cycle of stability in the
surface layer over land (unstable in daytime, stable during night).

Over the sea, for most areas of the oceans, we find the air one
or two degrees colder than the water, at least on the average. Of
course, there are larger variations of the air temperature caused
by advection of cooler or warmer air masses with synoptic or mesoscale systems. In certain parts of the oceans, the surface water
is cool, caused by upwelling of cooler water from deeper layers
to the surface. But since more solar energy is absorbed at the
surface than is radiated back by IR radiation (see Figure 7), commonly the sea surface is warmer than the air. When the air is warmer than the water, the air is cooled at the surface and a stable
stratification results. If the air is cooler than the water, the
air is warmed from below und unstable stratification results.

In an unstable situation, because a parcel removed from its equilibrium condition is accelerated, we must consider the stability
of density stratification over a greater height range. The rising
parcel is cooled adiabatically. It will continue to rise as long
as it remains warmer than its surroundings. With the cooling, the
saturation water vapor pressure decreases, and the relative humidity may eventually reach 100%. Even below 100%, part of the water
vapor is used to increase the water content of aerosol particles.
At 100% humidity further cooling leads to condensation of the excess water vapor into cloud droplets (or ice particles). The latent
heat is given to the air and yields a less rapid decrease of air
temperature during further ascent. This individual change of temperature of the ascending air is called the wet adiabatic lapse
rate. Consequently, ambient air which is stably stratified for
dry adiabatic ascent, may be unstably stratified for wet adiabatic
ascent, that is, of cloud air. This may give rise to fair weather
cumulus clouds or congested forms of cumuli, and in extreme cases
to thunder showers from deep cumulonimbi. The situation is depicted
in Figure 20. The process of vertical motion in congested clouds,
brought about by unstable stratification (both dry and wet), is
commonly called convection.

Different types of convection are found in fair weather situations, at cold fronts of synoptic systems, and at squall lines.
Also satellite photographs show convection to be a very common
feature over the world oceans, especially in the cold air at the
rear of depressions. They also show that convection often possesses
horizontal patterns, for example, in the form of hexagonal cells. A
pattern, which typically has a smaller space scale and can be observed from below, is formed by the so-called "cloud streets".
These are rows of low cumulus clouds. It is assumed that cloud
streets indicate roll motions in the boundary layer of the atmosphere. The motion is presumably in the form of helical rolls
oriented along (or nearly along) the wind direction, and reaching from the surface to a capping inversion (Figure 21).

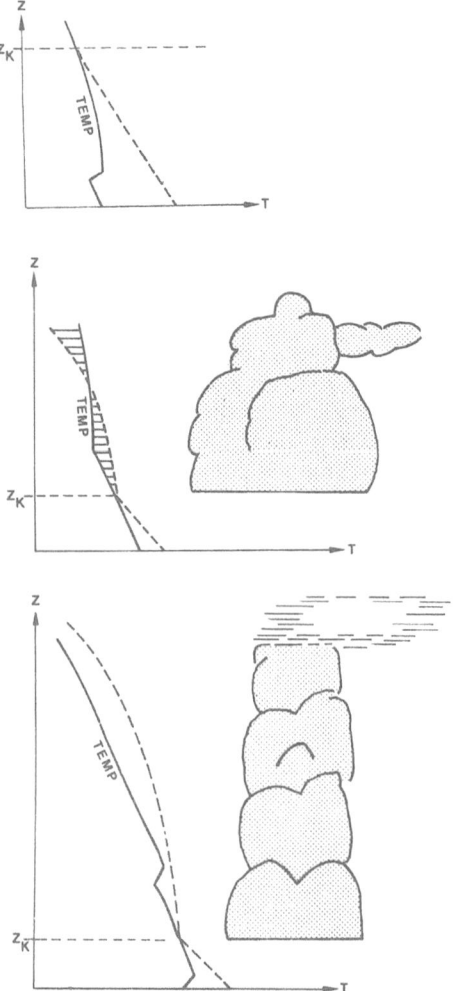

Figure 20. Different conditions of convective cloud
formation. z_K is the 'lifting condensation level'. Full
line is the temperature-height curve of the ambient air
as measured by a radiosonde ascent. Dashed line is the
individual temperature change along dry and wet adiabats.
In the upper graph, a parcel lifted from the ground with
the indicated temperature would not reach condensation.
In the middle graph, a lifted parcel would change its
temperature following the dry adiabat up to the condensation level and then following the wet adiabat until
it is stopped by the inversion. A shallow cumulus layer
would result. In the lower graph, a lifted parcel, which
has reached the condensation level, will (on its wet
adiabatic ascent) remain warmer than the ambient air

and be accelerated upwards. Deep cumulus convection
results. In the latter two cases, the stratification
above the condensation level is called conditionally
unstable, because the situation becomes unstable only
if condensation is reached.

The upward branch may or may not be made visible by cloud streets.
These helical rolls are difficult to measure (because of their
scale), but where they exist, their vertical motions are efficient
in carrying admixtures from the surface to the top of the boundary
layer, or vice versa.

In convective clouds, the vertical motions within the clouds
are concentrated in funnels of higher speed. Updrafts reach 10 or
20 m/s in well developed cumuli. The rising air is diluted by entrainment of dryer air from the sides. With raining clouds, evaporation of rain leads to cooling of the air, so that cold, wet
air is accelerated downwards and spreads at the ground. Also
downdrafts are produced by the friction of falling rain. Between the clouds we also find slow sinking motion. This is often
evidenced by the clear sky between cumulus clouds. This downward
motion is small and difficult to measure, but must be there for
continuity reasons. In a field of convection the area fraction
covered with clouds in a radar picture is 5% or 10%. Hence the
sinking motion is by a factor of 10 or 20 smaller than the upward
motion. The sign of the net mass flux is determined by the large-
scale convergence or divergence.

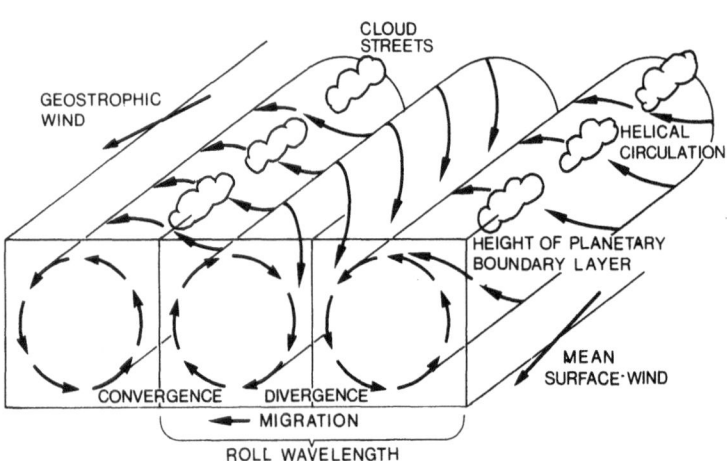

Figure 21.   Perspective schematic presentation of helical vortices and cloud streets. (Adapted from LeMone,
1972.)

6. TURBULENCE AND THE PLANETARY BOUNDARY LAYER

The different properties of the surface (land, sea), compared
to the air, have a strong influence on atmospheric flow and other
properties near the surface. The influence of the surface decreases with height. The layer where either in the motion field or in
the temperature structure the direct influence of the surface is
felt, is called the Planetary Boundary Layer (PBL) of the atmosphere. The Planetary Boundary Layer is most often defined in
terms of the influence of friction. In the free atmosphere, to a
good approximation, geostrophic equilibrium prevails; that is,
there is equilibrium between pressure gradient and Coriolis acceleration, and the flow is parallel to the isobars. Near the surface friction is effective as a third acceleration, and the flow
deviates from the isobars towards low pressure. The frictional influence and hence the angular deviation (cross-isobar angle) decreases with height until geostrophic equilibrium is reached
(Figure 22).

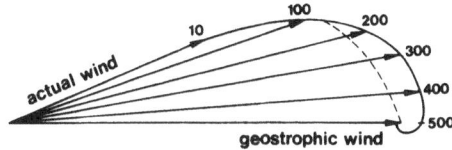

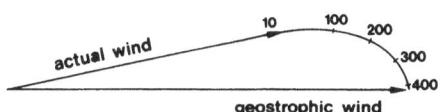

Figure 22. Wind veering in the planetary boundary layer of the atmosphere. The arrows represent wind vectors
at the indicated heights. The tips of the arrows, as a
function of height, form a spiral curve. Under simpli-
fying assumptions, this is a logarithmic spiral, called
the Ekman spiral. The dashed line is more typical for
observations. The upper graph is for conditions over
land, the lower one for sea. Actual spirals show a wide
range of variation from the given ones.

This level is taken as the height of the Planetary Boundary Layer.
It is usually 1000 m to 1500 m over the land (daytime) and 300
to 600 m over the sea.

In order to understand the processes in the Planetary Boundary
Layer, we discuss turbulent motion first. Any fluid (this includes

liquids and gases) may flow either in a laminar or turbulent state
of motion. These two states are best demonstrated with an experiment by Reynolds. In his study of flow through a pipe he made the
movements of the fluid visible by introducing a fine filament of
colour. He found two distinct types of flow: in "laminar" flow,
the filament formed a straight line, indicating that the flow was
layered. With higher speed, the filament became wavy and broke up,
the dye being mixed over the entire fluid. This second type of
fluid motion is called turbulent. It is a characteristic of turbulent flow that irregular motions in all three directions are
superimposed on the mean flow. These fluctuating velocity components mix the fluid. If there exists a gradient of a certain property in the fluid, the mixing can transport the property down its
gradient.

Turbulence can be characterized by the kinetic energy of the
turbulent velocity components. The energy of turbulence is taken
from the mean flow (which is driven by the horizontal pressure
gradients). The friction at the surface, for example, produces
a shear flow, that is, flow with a vertically changing wind vector.
This shear is instrumental in converting mean flow energy into
turbulent kinetic energy. The turbulence intensity is strongly
influenced by the ambient density stratification. In unstable
stratification, vertical motions are accelerated and receive energy from the density stratification. With stable conditions, on the
other hand, vertical motions induced by friction at the surface
must work against the density stratification. The work is taken
from the energy of turbulence and hence, under strongly stable
stratification, turbulence may decay or can not exist. The effects
of stability are often measured in terms of a Richardson-Number:

$$Ri = \frac{g}{T} \cdot \frac{\frac{\partial \theta}{\partial z}}{(\frac{\partial u}{\partial z})^2} \tag{11}$$

Instead of the Ri-number, other dimensionless numbers are used
with slightly different definitions, e.g. z/L, where L is the
Monin-Obukhov length (see e.g. Haugen, 1973). The Ri-number and
similar numbers measure the rate of work done by or against the
buoyant forces versus the rate of shear-production of turbulent
energy (unstable, Ri < 0; stable, Ri > 0).

Except for a very thin layer (a few millimeters deep) at the
surface, where the influence of the molecular viscosity is felt,
the motions in the atmosphere are always turbulent. Turbulence is
very active as a transporting agent. This is seen when, by analogy
to the molecular transport, the turbulent transport is described
by aid of an eddy diffusivity K

$$F = - \rho K \frac{\partial m_r}{\partial z} \tag{12}$$

where F is the flux of a certain property and m_r is its mixing
ratio $|kg/kg|$. A typical value of K (except near the surface) is
5 m^2/s. This is by a factor of 10^3 or 10^4 higher than the molecular diffusivities in air (see e.g. Table 3). Equation (12) is
applicable only for such properties which remain unchanged with
vertical motions. It can be derived in various ways, e.g. by di-
mensional analysis. It is usually a good description in fully turbulent flows such as found in the atmospheric surface layer.

Aside from the formal analogy of molecular and eddy diffusivity,
the physical processes of molecular and turbulent transport are
quite different. Molecular diffusivities depend on the property
exchanged, on the fluid in which the transport takes place, and
on the temperature of the fluid. The eddy diffusivity is independent
of temperature and more or less the same for all exchanged properties, but is highly dependent on the intensity of turbulence. This
means that the eddy diffusivity is proportional to wind speed
(more exactly, to the friction velocity, see below), modified by
stability, and is dependent on the distance from the surface (the
eddy size is limited near the surface for continuity reasons, and
hence the turbulent transport is less efficient). The typical
picture (see Figure 23) is a linear increase of the diffusivity
with height near the surface until 50 to 150 m, followed by a
decrease to a small (but non zero) value in the free atmosphere.

The description of eddy motion and transport processes in the
Planetary Boundary Layer is quite difficult. From the defining
equation of eddy diffusivity (12) it would seem, that only the
mean vertical gradient and the eddy diffusivity are needed in order to calculate the flux. But the eddy diffusivity is highly
variable, since it depends on the physics of the processes in the
Planetary Boundary Layer. This is true of other descriptions of
the Planetary Boundary Layer too. Consider the mechanical generation of turbulence. This depends on the friction at the surface
and on the variation of the wind speed and direction with height,
the wind shear. Surface friction can be characterized by a roughness length or a drag coefficient (which to a first approximation
depend only on surface geometry). The wind shear is brought about
by turbulent mixing and is variable; it is also influenced by the
shear of the geostrophic wind, that is the so-called thermal wind.

The most important parameter to influence the turbulent transport and hence the eddy diffusivity in the Planetary Boundary
Layer is the stability of density stratification, as measured e.g.
by the Richardson-Number. Even with small deviations from neutral
stability there is a distinct difference in the turbulent processes between stable and unstable stratification. On the stable side
the destruction of kinetic energy, by work against buoyancy, damps

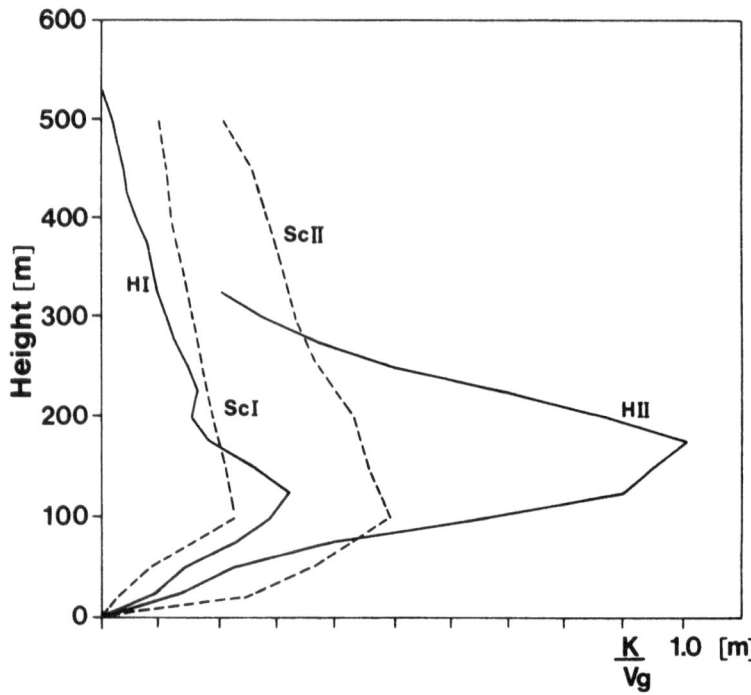

Figure 23. Variations of eddy diffusivity with height. Since the eddy diffusivity increases with wind speed, it is here normalised with the speed of the geostrophic wind. Examples are from pilot-balloon-derived wind profiles at the Scilly and Helgoland islands, so the diffusivities are for momentum (eddy viscosities) in the example. (Adapted from Hasse, 1968.)

turbulence. Hence transports are inefficient under stable conditions. Inversions (that are layers of temperature increase with height), therefore, are barriers for turbulent (or diffusive) transport of atmospheric admixtures. On the other side, in unstable conditions, the buoyancy induced vertical motions enhance turbulence. In the extreme, in strongly unstable conditions, buoyancy-driven convection dominates the mechanically generated turbulence. Near the surface, turbulence(in the sense of irregular, random motions) will still exist. But at some distance from the surface, the vertical motions become organized, and the concepts of eddy diffusivity and transport by turbulence fails. One finds constant potential temperature in the ascending and in the descending air so that the vertical temperature gradient vanishes. Hence from (12) zero vertical flux would be calculated, although there is strong transport by convective motions. In convective conditions we, therefore, usually find a surface layer with non-zero gradients

of temperature and wind speed, up to a few tens of meters, with a
well-mixed layer above (constant potential temperature, no variation of wind speed and direction).

In the simplest case of atmospheric boundary layer theory, in
neutral density stratification, the height of the boundary layer
is determined by the friction at the surface and is proportional
to the wind velocity (e.g., in the Ekman spiral). In the stable
situation, the boundary layer is compressed to the height range,
where frictionally induced turbulence can overcome buoyant damping. In the unstable case, the height of the boundary layer is the
height of the unstable or well-mixed layer. Unstable conditions
exist only near the ground. In the free atmosphere the stratification is stable, the potential temperature increases with height.

Over land, the height of the temperature inversion capping
the unstable boundary layer is mainly determined by the sensible
heat flux at the surface and can be estimated fairly well (Driedonks, 1982). Over the sea, the height of the unstable boundary
layer is determined not only by the heat flux at the surface, but
also by large-scale divergence or convergence and radiational cooling.

7. INTERFACE DYNAMICS

Within the Planetary Boundary Layer, the layer adjacent to
the surface is usually referred to as the surface layer, or constant flux layer, or Prandtl layer. This layer is typically of
order 20 m in height. The fluxes of heat and momentum are not
really constant within this layer, but a good description may be
obtained by assuming constant fluxes. Due to its moderate height
the time constant of this layer is of order a few minutes, and
horizontal homogeneity generally is a good assumption. Processes
in this layer are essentially influenced by the turbulent character of the flow. Since turbulent energy is produced by friction,
the roughness of the surface is one important parameter. Again,
turbulent energy is influenced by stability. We will discuss the
case of neutral stability first and consider the influence of
stability separately. We will discuss flow at a solid, level surface before we proceed to a fluid interface like the sea surface.

7.1 Flow at a solid, level surface

The model envisages a decreasing diameter of eddies as the
surface is approached. Hence an eddy scale-length or "mixing
length" is defined, which is proportional to the distance z from
the wall, say κz. Assuming a constant flux of momentum, given by
the shear stress τ at the surface, the vertical profile of the

modulus of the mean wind vector (the wind speed) u becomes

$$\frac{\partial u}{\partial z} = \frac{u_*}{\kappa z} \tag{13}$$

where u_* is called the "friction velocity" and is defined by

$$\tau = \rho_a \, u_{*a}^2 \tag{14}$$

τ is the modulus of the horizontal shear stress vector, the direction of which is in the mean wind direction. Integrating along z, which we assume to be perpendicular to the wall, we obtain

$$u(z) - u(z_o) = \frac{u_*}{\kappa} \ln \frac{z}{z_o} \tag{15}$$

This is called the logarithmic wind profile, applicable for neutral stability. Here the integration is extended to a variable height z. Usually it is assumed that there is no slip at the surface, $u(z_o) = 0$ and the integration constant z_o so-defined is called the 'roughness length'. z_o is a measure of the surface roughness. When Prandtl introduced this concept, he specifically called it a trick, which allows the logarithmic profile to be extended right to the surface. This implies that the eddy motions decrease in size in proportion to z right down to the wall. This is a good description as long as z is large compared to z_o, say by a factor of 100.

At the "wall" (which for the present we take to be a fixed, flat, but rough wall) the eddies are hindered in their movements perpendicular to the wall by its presence. Due to continuity constraints, the transport by eddies decreases more rapidly than linearly as the wall is approached. The decreasing energy of turbulence close to the wall has led to the definition of a "laminar" or better "viscous sublayer", where the fluid viscosity becomes important in the dynamics and the flow is dominated by viscous forces.

Experimental evidence (Figure 24) shows the typical profile of laminar flow near the wall, a transition region, and at larger heights the logarithmic profile of turbulent flow. The conceptional model is of a layer with decreasing turbulent transport as the wall is approached. Molecular transport adds to turbulent transport, since movement of the molecules takes place anyway, independent of additional eddy motions. In the transition region, molecular and eddy transport are of the same order, while with increasing distance from the wall eddy transport takes over.

Such continuity and boundary condition type arguments give only a gross picture. It is not really known how the flow behaves

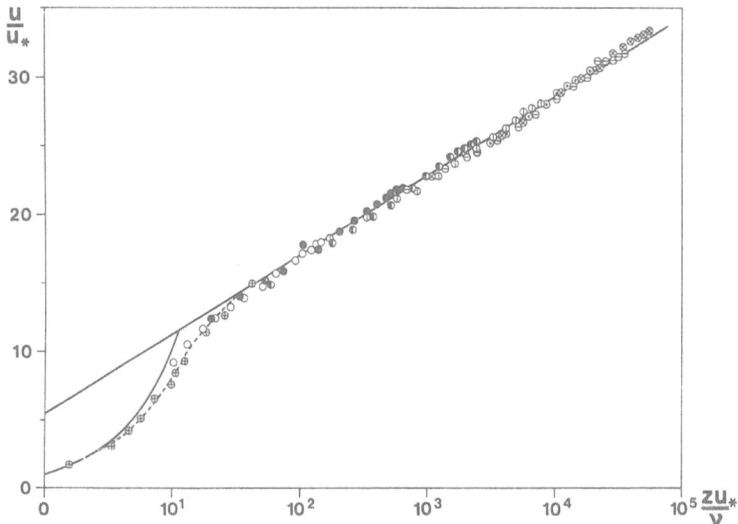

Figure 24. The velocity profile near the wall. Abscissa
(distance from the wall in dimensionless units) is logarithmic, hence the logarithmic profile is given by a
straight line. The linear velocity profile of laminar
flow on the left hand side is distorted according to
the logarithmic scale. (Adapted from Schlichting, 1965.)

near the wall. Does the scale and energy of the eddies decrease as
they approach the wall such that no eddy may really remove fluid
from the vicinity of the wall? Or is there just a decrease in
probability of an eddy reaching a given distance from the wall?
In principle, the power laws for parallel and perpendicular eddy
velocity components can be different, although there is a cross-
coupling between eddy velocity components effected by pressure
fluctuations. The question of how the eddy velocity components
and turbulent exchange decreases on approaching the wall is of
considerable interest, since molecular transport is so much more
ineffective than turbulent transport, especially for heat and gases with small molecular diffusivities. A review of power laws near
the wall is given by Monin and Yaglom (1965, Vol. I, Sec. 5.3 and
5.7).

The roughness in wind tunnel work is customarily produced by
gluing sand of a certain grain size to the wall or by determining
an "equivalent sand-grain size". The thickness of the viscous sublayer depends on the kinematic viscosity ν and is proportional to
ν/u_*. The transition from laminar to turbulent flow occurs at a
height of about 11 ν/u_*.

If the roughness elements are smaller than the thickness of

the viscous sublayer, the roughness elements do not influence the flow and the integration constant z_o may be replaced by a length proportional to ν/u_*:

$$u(z) = \frac{u_*}{\kappa} \ln \left(\frac{9u_* z}{\nu}\right) \tag{16}$$

This flow is called hydrodynamically smooth flow, while in the alternative case the flow is called hydrodynamically rough.

7.2 Flow at a fluid interface

In the case of a moving interface such as the sea surface, the definitions of smooth and rough flow are not directly applicable. In the sense used in physics, both gas and water are fluids, and the atmosphere-ocean interface can be seen as an internal interface, with waves propagating coherently on both sides of the interface (see e.g. Kraus, 1972). Waves and surface currents take up and transfer momentum and hence the momentum transfer is certainly different from the case of a solid wall.

Despite this, it can in practice be assumed that many of the concepts from laboratory work on smooth and rough surfaces can be applied at the mobile air-sea interface. The arguments are as follows. For a turbulent eddy approaching the interface, movements perpendicular to the interface are restricted as they are on approach to a solid wall. This is so because to deform the interface, work against gravity and surface tension would be necessary, and this work is much larger than the kinetic energy of turbulent motion of a small fluid parcel. Although the boundary conditions at a liquid/gas interface are different, the same power law is expected to hold as at a solid wall (Hasse and Liss, 1980).

A complication exists, however, if waves are present. The above mentioned derivation is for a flat, fluid interface. As long as the wave scales are large compared to the typical thickness of the viscous sublayer (a few millimeters or fractions thereof), the wave induced curvature of the sea surface is unimportant. This is not so in the case of capillary waves, whose radius of curvature may have the same order as the thickness of the viscous sublayer. Unfortunately, both the life cycle of capillary waves and wave-induced-near-interface-flows and their influence in gas exchange at the sea surface are not well understood.

7.3 Stability effects in the surface layer

Micrometeorologists have expended considerable efforts to determine the influence of stability on the fluxes of momentum, heat and water vapor in the surface layer. The stability dependence of turbulence-controlled fluxes is dealt with by similarity theory, often called "Monin-Obukhov similarity". In short, similarity theory

provides a relationship between fluxes and gradients of the transported property in the surface layer. This relationship depends
on a dimensionless stability parameter like the Richardson-number
or z/L, where L is the Monin-Obukhov length. A review is given in
Haugen (1973). Since the stability effects become important only
above the region of wave influence, stability functions determined
over land are applicable also at sea. It can be assumed that surface layer stability is not important for gas or particle exchange.
Stability modifies the turbulent energy and hence the turbulent
transport in the surface layer (or higher layers) only. Transport in the viscous sublayer or near to it is not influenced
directly. Near the surface, turbulence is mainly generated mechanically; the influence of stability becomes increasingly important
only above a height of a meter or so. There is an indirect influence of stability on the transports through the viscous sublayer,
since the sublayer thickness is inversely proportional to the
friction velocity. The latter depends on surface layer stability,
but this variation is of order 10% only if parameterization (see
below) is made in terms of surface layer variables.

7.4   Parameterization

The momentum flux at the surface can be parameterized reason-
ably well (with a scatter of 20% and a probable uncertainty of
50% at higher wind speeds) with aid of a drag coefficient C_D
(Figure 25) and the mean wind speed u:

$$\tau = C_D \, \rho_a \, u^2 \qquad\qquad (17)$$

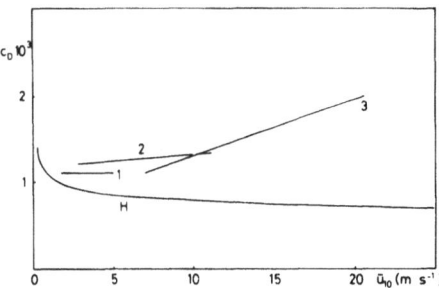

Figure 25. Roughness of the sea surface. Comparison between measured drag coefficients C_D (numbered lines)
and hydrodynamical smooth flow (curve marked H). The im-
plication is that the increase of momentum flux, compared to smooth flow, is brought about by form drag of
gravity waves and ripples. The average value at moderate
wind speeds is fairly well known both from direct and
profile measurements, while the value at high wind
speeds is based on few direct measurements. (After Hasse
and Liss, 1980.)

With (14) the thickness of the viscous sublayer in air can be determined. (The equation is written for the modulus of the horizontal stress component only, which enters in u_*.)

It is also possible to relate the surface stress to the geostrophic wind speed V_g, that is, to the surface pressure field.
This is done either by boundary layer modelling or by a similarity
theory known as "resistance law". A geostrophic drag coefficient
C_g is defined by

$$\tau = C_g \, \rho_a \, V_g^2 \tag{18}$$

The geostrophic drag coefficient depends on all processes which influence turbulence and convection in the Planetary Boundary Layer,
and is consequently quite variable. With the same neglect of detailed description of processes it is also possible to use a surface to geostrophic wind speed relationship (Figure 26) together
with the surface layer parameterization (17).

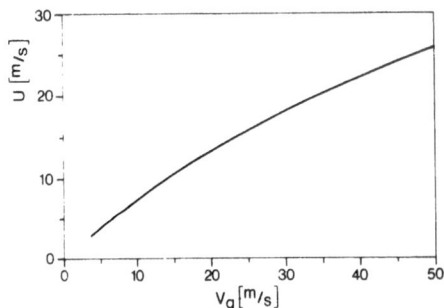

Figure 26. Surface to geostrophic wind relationship
from observations in the German Bight. (Adapted from
Luthardt and Hasse, 1982.)

The viscous sublayer thickness is determined by aid of the
friction velocity u_* in the respective media. The parameterizations
(17,18) are for the atmosphere. The friction velocity in the sea
is obtained by assuming constant flux of momentum across the interface:

$$\rho_a u_{*a}^2 = \rho_s u_{*s}^2 \tag{19}$$

In papers on gas and particle exchange usually a deposition
(or transfer or piston) velocity is used, which has dimension of a
velocity and hence will probably be proportional to some characteristic velocity of the problem (e.g. friction velocity). In micrometeorology,however,use of a dimensionless transfer coefficient

is preferred, like the drag coefficient (17). Take for example
the water vapor flux (evaporation) E. With q specific humidity,
and Δq the difference of specific humidity between air and sea
surface,

$$E = k_a \, \rho_a \, \Delta q \tag{20}$$

$$E = C_E \, \rho_a \, u\Delta q \tag{21}$$

Hence, from $k_a = C_E u$, micrometeorological results   (see e.g. Kraus,
1972) can be used to obtain deposition velocities (C_E is about
1.3×10^{-3}). The use of dimensionless transfer coefficients like
C_D , C_H , C_E has the advantage that the rough proportionality of
turbulent transports to mean wind speed is taken care of and the
transfer coefficients are less variable.

8. PRACTICAL ASPECTS OF ATMOSPHERIC TRANSPORTS

In the preceding sections atmospheric motions of different
scales have been discussed. Since the atmospheric processes are
very complex, for an understanding of meteorological conditions
for air/sea exchange of gases and particles it will often be
necessary to simplify and mainly consider processes at the scale
of interest. There is a large disparity between the horizontal
and vertical extent of the atmosphere and, similarly, in the horizontal and vertical components of atmospheric circulations, caused
by the different magnitudes of Coriolis acceleration ($\sim 10^{-3} m/s^2$)
and acceleration of gravity ($10 \, m/s^2$). Hence, it simplifies discussion if we deal with horizontal and vertical transports separately, although de facto horizontal and vertical motions are always coupled.

8.1 Vertical transports

We have already discussed some processes, which provide vertical upward transports: The mixing by turbulence causes vertical
transports if there is a gradient of a certain property. This
transport is effective only in the surface layer and the Planetary
Boundary Layer. If the source of a property is at the surface,
turbulence helps to transfer this property from the surface into
the lower troposhere, where other mechanisms may take over. Convection can be quite effective in transporting properties into
higher layers of the atmosphere. Note that with the diurnal variation of insolation over land there is enhanced turbulence and convection during daytime and suppressed turbulence and convection
during nighttime, with decoupled larger scale flow above the
nighttime inversion. A third mechanism producing upward transports

is the rising motion in midlatitude and tropical disturbances.
Additionally, there are mesoscale circulations, which may be important in given circumstances and areas.

Downward transports are different from upward transports,
due to different energetics of upward and downward motions. Turbulence mixes and provides downward transports if there exists a
gradient. Convection provides for downward transports both by
downdrafts and by the slow sinking motion in between clouds. Sinking motion in the high pressure cells is slow. If it were not
for gravitational settling and wet removal, the downward transport
of any admixtures would be slower than the upward transport.

In our picture vertical fluxes of gases or particles pass
through a series of layers: free atmosphere, planetary boundary
layer, surface layer, viscous sublayer in the air, interface, viscous sublayer in the ocean, mixed layer, deep ocean. These layers
have different physical properties and consequently different
transport resistances or conductivities. Depending on the property
being transported, sources and sinks may be in any of these layers,
while in the intermediate layers the fluxes are more or less constant. For a constant flux, it is unimportant in which part of
the path it is determined. In the case of exchange of gases and
particles, the viscous sublayers in the atmosphere and/or ocean
have the highest resistances (except for gravitational settling and
wet removal, which form a bypass). Hence the processes in and
near this layer and its thickness will be decisive.

There are two conclusions, which we can draw for those substances with the highest transport resistance in the viscous sublayer. First, the gradients above the viscous sublayer will be
fairly small and an attempt to determine fluxes by the gradient
method almost surely must fail. Since the admixtures are well
mixed in the surface and Planetary Boundary Layer, the eddy corre-
lation technique also will not work. The correlations between the
fluctuations of the admixture and the vertical velocity component
will be small and side effects become dominant (see Webb, 1982).
Second, the turbulent exchange coefficient K is probably nearly
the same for all passive admixtures. The surface layer has a time
constant of a few minutes. This is the time needed to reestablish
an equilibrium profile after a disturbance of the profile occurred.
The time constant of the Planetary Boundary Layer is of order a
few hours. Hence, with an average advection speed it can be estimated that a fetch of order 100 km is sufficient to establish an
equilibrium profile of admixtures in the Planetary Boundary Layer.

8.2 Horizontal transports

The atmosphere possesses a large spectrum of motions from
planetary waves, synoptic scale disturbances, mesoscale circula-

Residence time	Mechanism	Methods
Years	Stratospheric transport. Interhemispheric exchange.	Analysis of nuclear test products and atmospheric contaminants.
Months	Tropospheric transport.	Statistics of trajectories, e.g. from numerical models possible.
Days	Advection with synoptic disturbances and larger scale tropospheric flow.	Trajectory analysis from weather maps or models.
24 hours	Advection by synoptic and local wind.	Mesoscale models available, but wind systems depend strongly on orography.
3 hours	Advection by local winds, diffusion by turbulence.	Stability dependent diffusion models.

Table 5.   Transport classification.

tions, to turbulent fluctuations. Which of these scales are important will depend on the atmospheric residence time of the gas/ particle in question. A coarse classification is given in Table 5. The columns "residence time" and "mechanism" are not really independent. For example, the troposphere is turned over by weather systems. Thus, contaminants removed by contact at the sea (or land) surface are depleted from the entire troposphere. In contrast, the stratosphere is stably stratified, with less vertical motion and ineffective vertical diffusion. Hence, long-lived contaminants follow the stratospheric motions for a longer time. Some few explanatory remarks may be added:

(i) Trajectory analysis is most easily performed from weather-service synoptic maps, which commonly are available as contour lines (topography) of height of selected pressures, say 500 mb pressure. Winds are then given by the slope of the heights of pressure levels, and trajectories are calculated as if the identified air parcels would remain at these pressure levels. But if we assume adiabatic behaviour of an air parcel, we have seen from the first law of thermodynamics that in a stable atmosphere a parcel would remain at the level of its potential temperature. If we move the parcel in predominantly horizontal direction, it would tend to follow the surface of its equilibrium potential temperature. Such surfaces are called isentropic surfaces. They may be inclined to the pressure surfaces, which usually are taken as calculation levels in numerical models. After some time, the parcel will reach some other pressure level and have a different wind direction from the one determined at the original pressure level. In principle, since the potential temperature (above the Planetary Boundary Layer) increases with height, the vertical co-ordinate (i.e. pressure levels) could be replaced by isentropic surfaces (see Danielson, 1974).

Complications arise since an air parcel will remain on its isentropic surface only under adiabatic conditions. On the time scales of a day, diabatic processes become effective, e.g. temp-erature changes caused by radiation and evaporation/condensation. The change of potential temperature could in principle be calcula-ted, and the parcel moved to a revised isentropic level. But whether trajectories are determined on pressure levels or isen-tropic surfaces, the difficulty remains: either one needs vertical movements explicitly (to select the appropriate pressure level), or one needs diabatic heating by condensation, which is modelled from vertical movements. Also, radiational heating and cooling depend very much on cloud cover, type, and height distribution, and these depend on vertical motions. At present, vertical motions are poorly determined in numerical modelling.

Additionally, trajectory analysis has limitations resulting from the large range of scales of atmospheric motions. A cluster of parcels released together may be separated by diffusion and the parcels can then be included in different convective or mes-oscale motions. After some time, the parcels will have quite different pathways. This would be so even if numerical models were sufficiently detailed and exact. In fact, additional uncer-tainties are added through the coarse resolution either of numeri-cal models or observational networks and observing frequencies (Sykes and Hatton, 1976; Pack et al., 1978).

(ii) Mesoscale transport would take place with land-sea breeze systems. Since the system reverses between day and night, one might suspect that it would bring no net transport. In general, this is not true even if sea and land breezes are of equal strength. Consider a source over land near the surface so that the transport would be in the lower branch of the land/sea breeze system. During night-time, contaminants would be carried towards the sea. The same air might return to the land in daytime. If the residence time is of order 12 or 24 hours, some of the contaminants may be deposited at the sea surface. Hence there will be a net transport of conta-minants from the land to sea. In general, because of the consider-able variability of shoreline orography, there is probably no meso-scale model suitable to consider all actual cases. Hence,if it is necessary to consider actual fluxes at a given place, it is advis-able to investigate locally for the typical flow pattern.

(iii) Diffusion-models for local dispersion of gases or suspended matter are in wide use for different types of sources(point, area) and operations (continuous, burst). Note that the commonly used Pasquill-classes or similar stability- or diffusion-categories have been developed for use over land(e.g. Pasquill, 1962, p.209). The input parameters, which are used to characterize the radiation balance and hence (implicitly) the atmospheric stability over land, are not appropriate over sea. Because of different thermal properties, the stability over land is determined by the radiation balance, while over sea, stability is mainly influenced by advection of cold or warm air or water, as represented by the air-sea temperature difference. In order to translate the Pasquill classes from land to sea, the following scheme seems feasible (Weber and Hasse, 1985) : The stability classification given in terms of radiation balance and hence vertical heat flux over land could be replaced by a parameterization of the sensible heat flux H at sea, e.g. by

$$H = C_H \cdot C_p \rho u \Delta T \tag{22}$$

where ΔT is the air minus sea potential temperature difference and the bulk transfer coefficient C_H is about 1.3×10^{-3}. In order to account for the different roughness of land and sea, the windspeeds given for land should be increased by a factor of 2 for use over sea.

GENERAL REFERENCES

Petterssen, S., 1969: Introduction to meteorology. Third Edition, McGraw-Hill, New York, 333 pp.

Wallace, J.M. and P.V. Hobbs, 1977: Atmospheric science. An introductory survey. Academic Press, New York, 467 pp.

Kraus, E.B., 1972: Atmosphere-ocean interaction. Clarendon Press, Oxford, 275 pp.

Haugen D.A.,(Editor)1973: Workshop on micrometeorology. American Met. Soc., Boston, 392 pp.

Nieuwstadt, F.T.M. and H. van Dop (Editors), 1982: Atmospheric turbulence and air pollution modelling. D. Reidel, Dordrecht, 358 pp.

Danielsen, E.F. and J.W. Deardorff, 1978: Modeling the atmospheric transport of pollutants and other substances from sources to the oceans. In: U.S. Nat. Acad. Sci., The tropospheric transport of pollutants and other substances to the oceans, pp. 25-52.

Eliassen, A., 1980: A review of long-range transport modeling. J. Appl. Meteorol. 19, 231-240.

Frenkiel, F.N. and R.E. Munn (Editors), 1974: Turbulent diffusion in environmental pollution. Adv. Geophys. 18 B, Academic Press, New York, 389 pp.

REFERENCES

Danielsen, E.F., 1974: Review of trajectory methods. In:Frenkiel
    and Munn (Editors), Turbulent diffusion and environmental
    pollution. Adv. Geophys. 18 B, Academic Press,New York,
    73-94.
Defant, A. and Fr. Defant, 1958: Physikalische Dynamik der Atmo-
    sphäre. Akad. Verlagsges., Frankfurt, 527 pp.
Driedonks, A.G.M., 1982: Models and observations of the growth of
    the atmospheric boundary layer . Boundary-Layer Meteorol.
    23, 283-306.
Hasse,L., 1968: Zur Bestimmung der vertikalen Transporte von Im-
    puls und fühlbarer Wärme in der wassernahen Luftschicht
    über See. Hamburger Geophys. Einzelschriften, 11, Cram de
    Gruyter Verlag, Hamburg, 70 pp.
Hasse, L. and P.S. Liss, 1980: Gas exchange across the air-sea
    interface. Tellus 32, 470-481.
Hasse, L. and H. Weber, 1985: On the conversion of Pasquill
    categories for use over sea. Boundary Layer Meteorol. 31,
    177-185.
Köppen, W., 1899: Grundlinien der maritimen Meteorologie.
    Niemeyer Verlag, Hamburg, 88 pp.
LeMone, M.A., 1972: The structure and dynamics of the horizontal
    roll vortices in the planetary boundary layer. Ph.D. thesis,
    Univ. Wash., Seattle, 128 pp.
List, R.J., 1958: Smithsonian meteorological tables. Sixth re-
    vised edition. Smithsonian Inst. Washington D.C., 527 pp.
Luthardt, H. and L. Hasse, 1982: The relationship between pressure
    field and surface wind in the German Bight area at high wind
    speeds. In: Sundermann and Lenz (Editors), North Sea
    Dynamics. Springer Verlag, Berlin, 340-348.
Monin, A.S. and A.M. Yaglom, 1965: Statistical fluid mechanics.
    Transl. from Russian, Cambridge, MIT Press, Vol. I, 1971,
    769 pp., Vol. II, 1975, 874 pp.
NAS, 1975: Understanding climate change. National Academy Press,
    Washington, D.C., 239 pp.
Newell, R.E., J.W. Kidson, D.G. Vincent and G.J. Boer, 1972: The
    general circulation of the tropical atmosphere and inter-
    actions with extratropical latitudes. MIT Press, Cambridge,
    Mass., Vol. 1, 258 pp., Vol. 2, 371 pp.
Pack, D.H., G.F. Ferber, J.L. Heffter, K. Telegadas, J.K. Angell,
    W.H. Hoecker and L. Machta, 1978: Meteorology of long-range
    transport. Atmos. Environ. 12, 425-444.
Pasquill, F., 1962: Atmospheric diffusion. Van Nostrand, London,
    297 pp.
Priestley, C.H.B., 1959: Turbulent transfer in the lower atmosphere.
    Univ. Chicago Press, Chicago, 130 pp.
Schlichting, H., 1965: Grenzschicht-Theorie. 5. Aufl., Braun,
    Karlsruhe, 736 pp., Engl. ed.: Boundary-layer theory, McGraw-
    Hill, New York.

Sykes, R.I. and L. Hatton, 1976: Computation of horizontal traj-
    ectories based on the surface geostrophic wind. Atmos.
    Environ. 10, 925-934.
Webb, E.K., 1982: On the correction of flux measurements for the
    effects of heat and water vapor transfer. Boundary-Layer
    Meteorol. 23, 251-254.

INTRODUCTORY PHYSICAL OCEANOGRAPHY

Fred Dobson

Bedford Institute of Oceanography
Dartmouth, N.S. B2Y 4A2
Canada

LARGE-SCALE OCEANOGRAPHY

The oceans are typically one thousand times broader than
their depth. One can infer from this that their dynamics ought
to be primarily two-dimensional, and this is in fact so, but
important deviations occur. Both horizontal and vertical circulations have vital roles to play, as we shall see. The principal
driving forces for the ocean are solar heating, the winds, and
the attraction of the moon.

Both ocean and atmosphere obey the same physical laws, including thermodynamics, but there are important differences
between them, which lead to differences in response of the two
systems to external forcing and in our techniques for observing
them.

The large-scale features of the tropospheric circulations in
the atmosphere - the Aleutian low, the Iceland low, the Azores
high, the subtropical convergence, the Hadley cells, and other
analagous features - set up wind stress (i.e. wind drag) patterns
on the surface of the sea. These stresses drive the large ocean
gyres by causing surfaces of constant density to tilt, which in
turn causes ocean currents to flow.

Whereas the atmosphere is free to move everywhere on the
earth's surface, being constrained only in the sense that the
mountain ranges exert more friction and to some extent direct the
air flow at low levels ("orographic" flow), the oceans are constrained by the land to move within fixed basins. The resulting
flow patterns in the atmosphere are roughly zonal (that is, along

53

latitude lines) and gyral (circular) in the oceans. Figure 1,
from Sverdrup et al. (1942), displays the major ocean currents in
February/March.

Whereas meteorologists can accurately measure the surface
wind speed and the surface pressure differences which cause geostrophic flows, oceanographers cannot. Even at the edges of the
continents, it is difficult to measure the sea surface elevation
well enough to compute the mean currents. The fundamental
problem is simple: the surface of the sea moves, both vertically
and horizontally, in ways not directly related to the local wind,
and the motions are so large they obscure the signal to be
measured.

This section will begin with some definitions, and then will
describe the fundamental dynamic and thermodynamic balances which
exist in various parts of the ocean. All are simplifications
from the full set of mathematical relations, describing fluid
flow, which are described in the Meteorology section. The
particular flow configurations they produce have been named after
their discoverers, hence the names like "Ekman" and "Sverdrup".
Taken together, the flows can be used to generate plausible
windplus density-driven ocean circulation models. Unfortunately,
although the models so far developed produce qualitatively
correct ocean circulation maps, none are completely quantitative,
and therein lies a lot of the excitement of theoretical ocean
dynamics.

Geostrophic, Baroclinic and Barotropic Motions

In the atmosphere, two important considerations make it
possible to measure or infer the three-dimensional velocity
field. First, the density field, the pressure field and the
velocity field of the atmosphere can be monitored routinely at
fixed locations. Second, the geostrophic relations (see
Meteorology section) specify that flow velocities must be (horizontal pressure gradient) (Coriolis parameter x density). In
the case of the atmosphere at mid-latitudes a 100 Pa/m pressure
difference over 1000 km produces a flow of 10 ms$^-$, which is
relatively easy to measure to an accuracy of, say, 10%. In large
areas of the ocean the corresponding flow rate is 0.1 ms$^-$, and
such currents can typically be measured to 10% only after
averaging over periods of several months to more than a year,
either with floats which follow water parcels or fixed current
meters. The data from both floats and current meters, averaged
over a year, have typical precisions of 0.01 ms$^-$. As a result,
oceanographers only very rarely know the absolute current at a
given location.

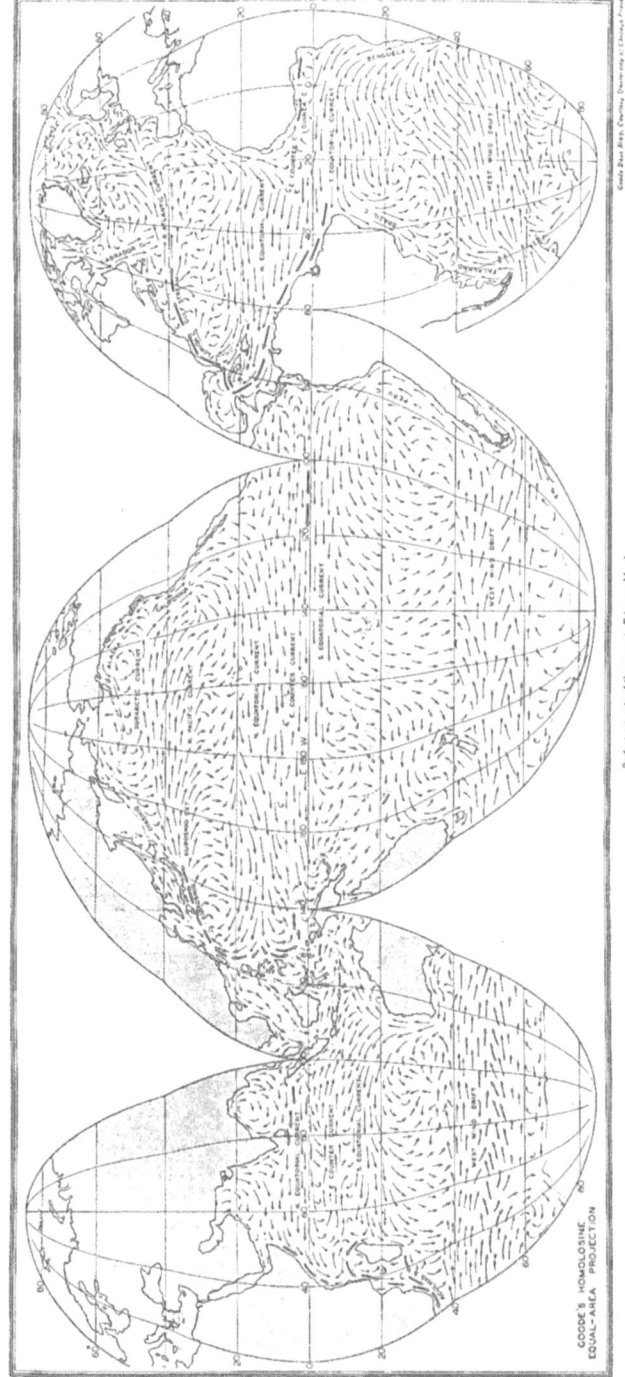

Figure 1. Chart of the ocean currents of the world in February-March. (The only major currents which change a significant amount with season are those of the Indian Ocean.) Much of the information shown was reconstructed from accumulated measurements of ship drift. (After Sverdrup, Johnson and Fleming, 1942.)

Setting direct current measurements aside, oceanographers
determine relative currents using the geostrophic relations, the
hydrostatic relation, and depth profiles of seawater density.
Typically a "depth of no motion" is assumed, usually in very deep
water where vertical pressure gradients are very uniform, although if it is available (as it might become if absolute satellite altimetry ever comes to pass) a "depth of known motion" is
of course superior. Horizontal gradients of density, vertically
integrated with hydrostatics to give pressure, then give the
geostrophic currents relative to that level, or the so-called
"geostrophic shear". A typical calculation is given in Pond and
Pickard (1983), p. 75 ff.

Such calculations, supplemented by routine measurements of
ship drift and inferences from property distributions (see
later), has been the principal tools by which oceanographers have
mapped the currents of the worlds oceans over the last 50 years.
The principal drawback is that the calculations give relative
currents only, (i.e. "baroclinic" motions resulting from non-
parallel pressure and density surfaces, that is, "isobars" and
"isopycnals"). As a result,oceanographers have had to rely on
inferential techniques for estimating the reference, or "barotropic" motions, which are due to a uniform tilt of both isobars
and isopycnals.

It should be noted that the definitions of "baroclinic" and
"barotropic" flow given here are not always held to by oceanographers. Sometimes "barotropic" is taken to mean "depth mean"
and "baroclinic" to mean "deviation from the depth mean". The
reference velocity used for baroclinic flows is generally that of
the deep water, but sometimes use is made of the surface velocity
or a velocity where measurements exist.

"Ekman" Transport

V.W. Ekman (1905) considered an infinitely deep and infinitely broad ocean (i.e. no bottom friction and no horizontal
boundaries), being acted on by a steady wind, with no pressure
gradients or density inhomogeneities. He allowed for friction at
the sea surface and within the system by vertical shear of the
current and used a shear friction coefficient independent of
depth. His solution (Figure 2) gave a spiral current, which at
the surface travelled at 45° to the right of the wind (cum sole)
in the Northern Hemisphere, in which the Coriolis force was
everywhere balanced by friction. The total mass transport (i.e.
integrated over the depth to which the wind-driven currents
penetrate-generally taken to be the depth of the mixed layer) is
found to be

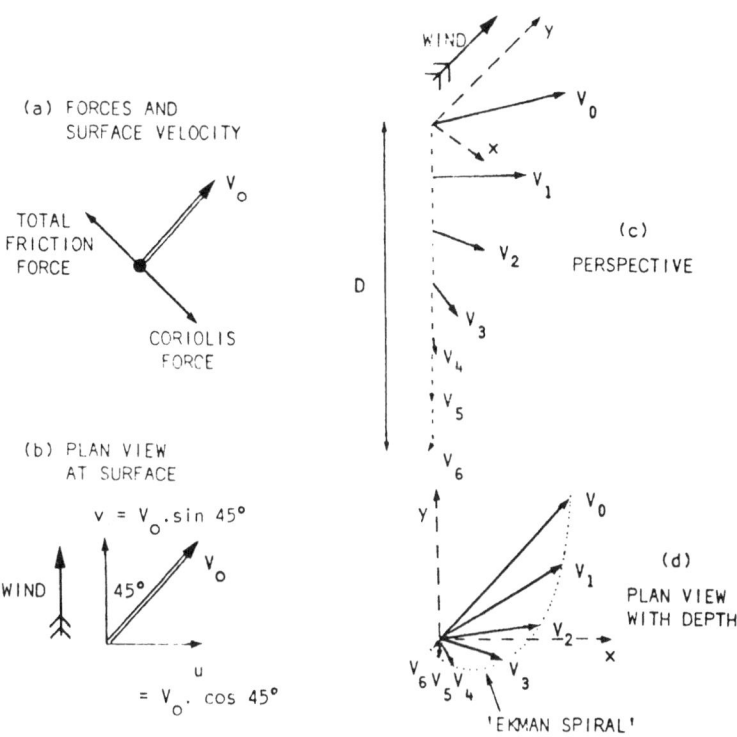

Figure 2. Ekman's solution (in the Northern hemisphere) for wind-driven currents, in which friction in the fluid is balanced by the Coriolis force. The surface current $\vec{V}_0$, is caused by the wind; the current is seen to decrease and turn clockwise with increasing depth. (After Pond and Pickard, 1983.)

$$M_{xE} = \tau_y/f \tag{1}$$

$$M_{yE} = -\tau_x/f \tag{2}$$

where $f = 2\Omega \sin \phi$ is the Coriolis parameter; Ω is the earth's angular rotation rate and ϕ is the latitude. Here τ_y and τ_x are the meridional (north-south) and zonal (east-west) components of the "wind stress", or the rate per unit area at which the wind imparts its horizontally-directed momentum to the sea surface. In the case of a steady current flowing over the sea bottom with friction, the "Ekman spiral" is also expected to apply, and has in fact been observed, although it is often obscured by other effects. Although (1) and (2) involve many simplifications, and probably do not represent physical reality, they produce quanta-tively useful estimates of wind-driven transports in the oceans.

"Sverdrup" Transport

Harald Sverdrup (1947) started with the linearized (i.e. no "advective" terms) basic equations, as did Ekman. He included pressure gradients but ignored horizontal friction. As did Ekman, he assumed his steady-state "ocean" was driven by a steady wind stress. He deviated from Ekman in allowing the wind stress to vary with latitude and longitude. Only the depth-integrated currents - the mass transports - were considered, and no lateral oceanic boundaries were permitted.

Sverdrup's equation was

$$M_y \, \partial f/\partial y = (\partial \tau_y/\partial x - \partial \tau_x/\partial y) \tag{3}$$

The derivative $\partial f/\partial y$ on the left hand side results from the commonly-used approximation

$$f = f_o + y\partial f/\partial y \equiv f_o + \beta y \tag{4}$$

where f_o is the value of f at the latitude of y. The right hand side of the equation is the vertical component of the "rota-tional" derivative of the wind stress, or curl. Thus (3) can be written

$$\beta \, M_y = \mathrm{curl}_z \, \vec{\tau} \tag{5}$$

This is known as the "Sverdrup equation" for the meridional oceanic transport. The "Sverdrup balance" is widely assumed to hold in the open ocean, away from western boundary currents (see

later). The zonal and meridional components of the mass transport, M_x and M_y, can be separated unambiguously into depth-
integrated "Ekman" and "geostrophic" components

$$\int_{-D}^{o} \rho\, U dz = M_x = M_{xE} + M_{xg} = (1/f)\, (\tau_y + \int_{-D}^{o} (\partial p / \partial y)\ dz)\quad(6)$$

$$\int_{-D}^{o} \rho\, V dz = M_y = M_{yE} + M_{yg} = (1/f)\, (-\tau_x + \int_{-D}^{o} (\partial p / \partial x)\ dz)\quad(7)$$

$$= \operatorname{curl}_z \vec{\tau} / \beta$$

Westward Intensification

The next step in understanding the ocean circulation came
with Henry Stommel's (1948) explanation of why the oceanic gyres
are stronger on the westward sides of the oceans ("westward
intensification"). This represented a large step forward in
dynamic oceanography, and understanding Stommel's explanation is
central to understanding ocean circulations in general.

The mean global wind patterns (Figure 3) show a reversal
in direction at mid-latitudes, from the high-latitude westerlies
to the low-latitude trades. In the Northern Hemisphere this
combination induces surface (Ekman) flows in the ocean at 90° to
the right of the winds, which converge at mid-latitudes. The
ocean, on a long time-scale, responds baroclinically, that is, by
changing its vertical density structure. The water does not
"pile up", but instead the lighter surface water pushes downwards
on the heavier water below the mid-depth permanent temperature
discontinuity known as the "main thermocline" (see the Mixed
Layer Section), forcing the deep water outwards. The response of
the ocean to the outwards acceleration is to form a geostrophic
gyre, clockwise in the Northern Hemisphere (see Figure 4). The
waters of the gyre, and in fact all fluids on a rotating plat-
form, such as the earth, are also subject to the constraint that
they must keep their total angular rotation rate, or spin, a constant (the actual rule is that they must keep total spin divided
by depth constant, but we ignore depth variations in this simplified argument). This means that on the eastern side of the gyre,
the equatorwards-travelling water, forced to gain clockwise spin
relative to its surroundings as it increases its distance from
the earth's axis of rotation, must acquire a counterclockwise
spin to keep its total spin constant. Similarly, polewards-
travelling water on the western side of the gyre has a tendency
to spin clockwise. The wind stress pattern at the same time

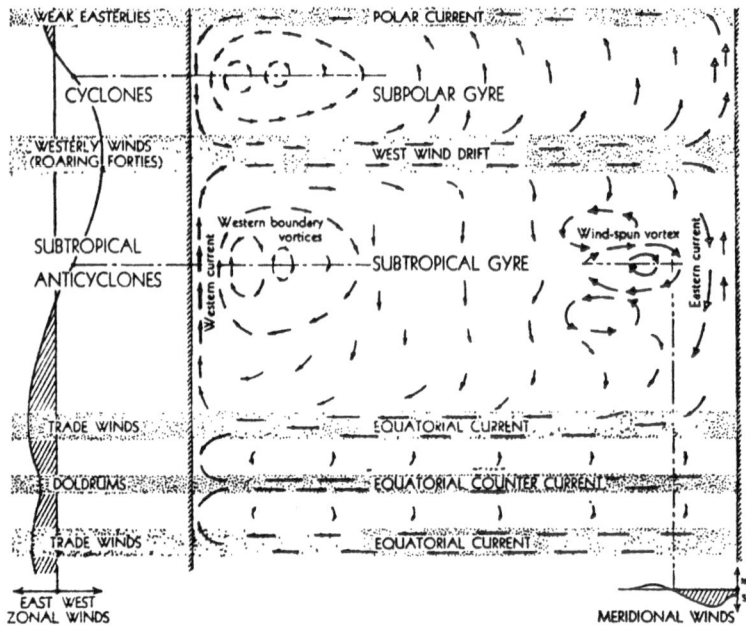

Figure 3. Schematic representation of the circulation in a
rectangular basin resulting from the principally east-west
(zonal) winds over the oceans (north-south (meridional) profile
of zonal winds at left). The nomenclature applies to either
hemisphere, but in the Southern Hemisphere the subpolar gyre is
replaced by the Antarctic Circumpolar Current (the "west wind
drift" shown in Figure 1). (After Munk, 1950.)

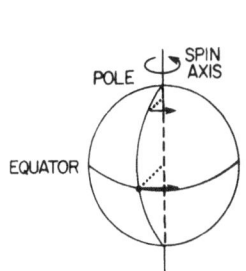

Intrinsic CCS (Counter ClockWise) angular momentum of fluid on earth's surface is zero at poles, maximum at equator: distance r from spin axis increases, equatorward.

Angular momentum

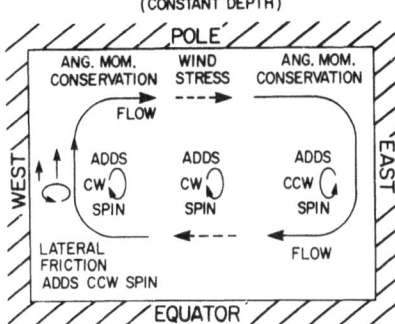

Angular momentum

Poleward-moving water conserves total angular momentum by spinning CW to offset CCW tendency caused by decreasing distance from spin axis.

Equatorward-moving water conserves total angular momentum by spinning CCW to offset CW tendency caused by increasing distance from spin axis.

Spins do not balance. Flow speeds up until lateral friction with boundary provides balancing CCW spin.

Spins balance, flow in equilibrium.

Figure 4: Westward Intensification. Schematic of the flow in a rectangular, flat-bottomed "ocean" on a rotating earth, induced by a wind stress pattern (dotted line) which is eastwards near the pole and westwards near the equator. To conserve total spin (vorticity) in the gyral flow, the polewards-moving water on the western side of the "ocean" must speed up until its CW spin is balanced by lateral friction with the boundary. The result is a circulation which is "Westward intensified". The Gulf Stream and the Kuroshio are examples of the effect.

induces everywhere a clockwise spin (in the Northern Hemisphere) to the surface waters. On the western side of the basin, the water gains clockwise spin from the wind and from its polewards motion, and, this leaves an imbalance. The flow on the western side of the ocean is forced to speed up; it intensifies until a balance exists between the clockwise spin gained from the wind and by polewards motion and lost by lateral friction against the western boundary. On the eastern side of the basin the spins induced by wind and equatorwards motion are opposite and roughly equal in magnitude, so a balance exists and no speed-up of the flow occurs.

The two most prominent examples of westward-intensified ocean currents are the Gulf Stream and the Kuroshio. Both have been extensively modelled; N.P. Fofonoff gives an up-to-date review of the present status of Gulf Stream models in Warren and Wunsch (1981). An article by G. Veronis in the same book provides an excellent theoretical account of large-scale ocean circulation dynamics. For a physically more complete, mathematically less complicated treatment, the book by Stommel (1960) provides first-class reading.

Physical Properties of Sea Water

Sea water density is defined by a rather complicated equation of state:

$$\rho(s,T,p) = a_1(s,T,p)s + a_2(s,T,p)T + a_3(s,T,p)p \qquad (8)$$

where a_1, a_2 and a_3 are empirically determined constants, which has been extensively tabulated. The present standard is presented as a set of algorithms giving the full equation of state and relating salinity to electrical conductivity, temperature and depth, in UNESCO (1981a,b). Before the advent of reliable conductivity-measuring techniques, the salinity was determined chemically by titration with silver nitrate. Density was obtained from temperature and salinity using the tables of Knudsen (1901).

The compressibility of water, although small, is not negligible in the ocean. The water in the deep ocean (below 4 km) is under a sufficiently heavy load that its in situ temperature actually increases with depth due to compression. To avoid problems in estimating the stability of the deep waters and in measuring the density of water moving over large ranges in depth, it is customary to use "potential" temperature, usually referred to as θ, and potential density, or sigma-theta. These are the temperature and density the water would have if decompressed from in situ to sea surface pressure without thermal contact with the surrounding water (that is, "adiabatically").

The formula to convert in situ temperature T to potential temperature θ is (see e.g. Phillips, 1977)

$$\theta = T - \int_{p_o}^{p} (\partial T / \partial p)_{S,s} \, dp \tag{9}$$

where p_o is the surface pressure and p that in situ, S is entropy and s is salinity (the subscripts mean the quantities mentioned are held constant). For potential density ρ_{pot}

$$\rho_{pot} = \rho - \int_{p_o}^{p} (\partial \rho / \partial p)_{S,s} \, dp \tag{10}$$

Water Property Distributions and Tracers

Oceanographers have for many years used distributions of various water properties and solutes to infer large-scale flows in the ocean (see, for example, the charts at the end of "The Oceans", by Sverdrup et al., 1942). The original technique was to map the large-scale distribution of a given "tracer", make some assumptions about how the tracer might change its concentration in situ with the passage of time, and then attempt to reconstruct the path from its source. Water "masses" are defined which retain some property, such as a particular shape on a plot of temperature versus salinity (T-s diagram), as they spread from the region where they are formed. Plots of one property versus another (T-s, s-O_2, O_2-Si, etc.) are much used by descriptive oceanographers as a means of characterizing water masses. Figure 5, from Gordon (1982) is a typical example of such a plot. With such diagrams mixing of two water masses with conservative characteristics (i.e. no processes exist which create or destroy the relevant properties) occurs on straight lines joining the two points defined by the two types. A "water type" is defined, and used here, as water with uniform temperature and salinity; that is, it forms a point on a T-s or θ-s diagram. "Thermostad", "halostad" and "pycnostad" refer to regions in the ocean with uniform temperature, salinity, or density.

Since the period of extensive nuclear weapons testing in the 1960's, many new radioactive tracers have become available. Since they were injected at the sea surface over a relatively short period and their half-lives are well-known, they can be used to delineate pathways by which water parcels travel in the ocean. The principal uncertainty is whether the water carrying the tracer has been carried there by currents (i.e. "advected") or the tracer has diffused into the region. The GEOSECS program

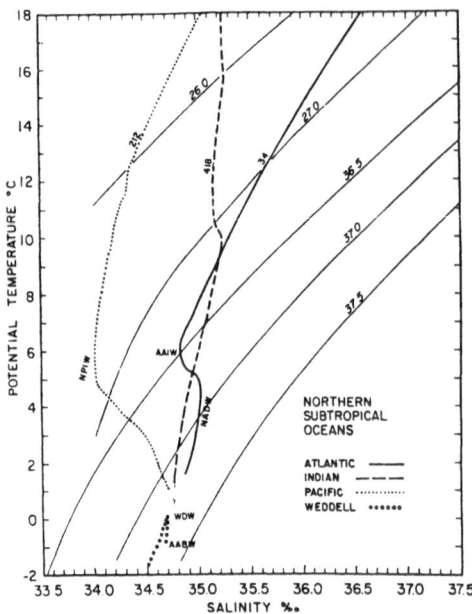

Figure 5. Temperature-salinity (T-s) diagram for the waters of
the northern subtropical oceans. (For a definition of potential
temperature see text, page 10.) The lines shown for the various
water masses are averaged over many observations. The thin
diagonal curves are constant-density lines, and the densities
given are 1000 (potential density in kg m^{-3} - 1.000), commonly
referred to as "sigma-theta". The numbers given beside each
water mass line are specific GEOSECS station numbers. NPIW is
"North Pacific Intermediate" water; AAI is "Antarctic Intermediate"; NAD is "North Atlantic Deep"; WD is "Weddell (Sea)
Deep"; and AAB is "Antarctic Bottom". (After Gordon, 1982.)

(Geochemical Ocean Sections Study) provided excellent baseline
surveys in the 1970's, and the TTO program (Transient Tracers in
the Ocean) is now providing additional data on the short halflife
tracers, to study the mechanisms by which large-scale oceanic
mixing occurs on climatic time scales. The reader interested in
pursuing the details should consult the article by W.S. Broecker
in Warren and Wunsch (1981), and Roether (1982).

Thermohaline Circulation

     Flows which are caused by the sinking of newly-formed, dense
water near the poles and subsequent upwards diffusion elsewhere
are called "thermohaline". Over most of the ocean the upper
mixed layer, with its relatively stable, almost frictionless
lower boundary (the so-called" permanent thermocline"), isolates
the deep waters from the direct influence of the driving forces
of solar radiation and wind. Therefore the mechanisms by which
the deep circulations are driven are quite different from those
which drive the surface flows. Only in a few places on the
globe, notably near the poles, are the deep waters not thus isolated. In the polar regions relatively warm, salty surface water
from lower latitudes encounters sea ice and intensely cold, dry
arctic air. Three processes occur: the warm, saline water loses
heat rapidly to the air through evaporation and direct conduction, and large volumes of sea ice are formed, releasing salt.
All three cause the density of the surface waters to increase;
these waters then sink, usually in small (<10 km diameter)
regions. Lateral mixing with surrounding water (often the surrounding water has its origins at higher latitudes, and is relatively cold and fresh) produces large amounts of water with
closely-defined temperature and salinity characteristics, which
then spreads out from the polar regions along the bottom of the
worlds' oceans. Examples of these deep waters are to be seen in
Figure 5, at the bottom of the θ-S curves. A clear example of
the way they flow out of the arctic regions is shown in Figure 6.

     To complete the thermohaline circulation, return "flows" are
needed of magnitude equal to the global-average rate of deep
water formation. Stommel and Aarons (1960) proposed the first
real model; how quantitatively correct it is remains unknown.
The sources of the oceans' deep waters are taken to be in small
regions near the poles, and the return flow is by a general slow
upward motion of the deep water. The latter accounts for the
presence of the permanent thermocline (a relatively sharp change
in temperature at intermediate depths - 300 to 1200 m - which is
found everywhere except at high latitudes), since it balances the
downwards diffusion of heat from the warm upper layer, which
tends to eliminate the thermocline. The mean horizontal flow in
the deep ocean is assumed to be geostrophic, and, as Stommel and

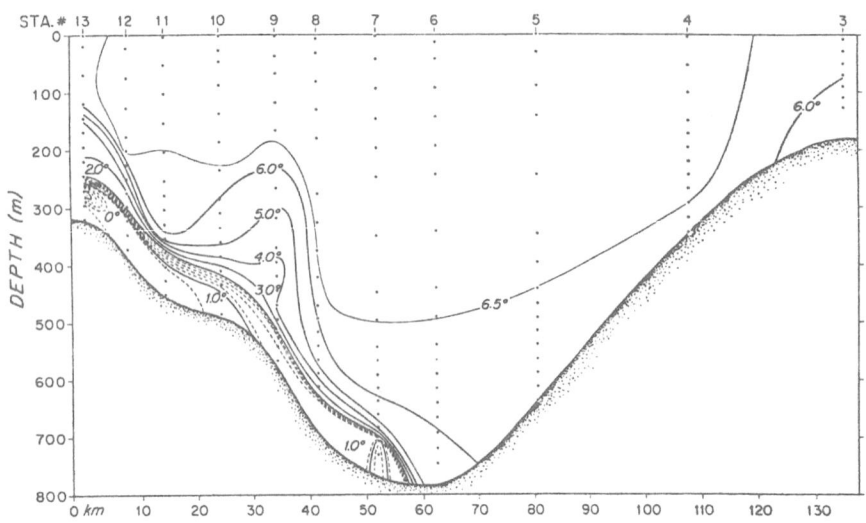

Figure 6. Temperature section across the Denmark Strait, looking
polewards (Lat. 65-66N). The cold water is flowing southwards
from the Norwegian Sea, hugging the western side of the Strait.
From CSS Hudson Cruise BI-02-67, January 1967. (After
Worthington, 1969).)

Aarons point out, this implies, contrary to intuition, that the thermohaline flows over most of the ocean must be polewards, towards their source. The only way to get the dense polar water south is to posit deep equatorward western boundary currents, which is what Stommel and Aarons did. The arguments leading to westward intensification of the thermohaline circulation are identical in spirit to those given in the preceding section; they are well summarized in the article on the deep circulation of the ocean by B.A. Warren in Warren and Wunsch (1981). Stommel's (1960) schematic of how things might work in the Atlantic is given in Figure 7.

It should be realized that the deep, density-driven flows are not understood quantitatively. Only rarely has the strength of the deep flows been measured well: the difficulties are enormous. The flows typically have large temporal and spatial variability (due partly to "barotropic" - i.e. depth-independent-components of mesoscale eddies, and partly to the deep waters' proclivity for following bottom contours), so that long-term (several-year) measurement programs are required to obtain useful statistics of the deep flows. And because the deep waters are so voluminous, very small mean flows can account for enormous transports of water. (In water 5 km deep, a 1000 km wide mean current of 0.5 cm s^{-1} would transport 25×10^6 m^3 s^{-1}, about the same as the mean Gulf Stream transport through the Florida Straits.)

The thermohaline circulation, as poorly known as it is, has recently become the subject of intense interest, partly because the deep waters are an enormous sink for anthropogenic increments in airborne CO_2, and partly because man wants to store nasty things with long half-lives on or in the sea bottom (If they are out of sight, can they really be put out of mind?). The GEOSECS and TTO programs were both designed to define the mean state of the deep waters and to estimate lifetimes of tracers injected in polar regions. The next phase of exploration, now being planned as part of the World Climate Research Program, will be to study the response of the deep circulations to perturbations in their driving processes. The reader with curiosity about the details would do well to read Stommel's (1960) book and the Stommel memorial volume by Warren and Wunsch (1981). The CO_2 problem is lucidly addressed by Broecker et al. (1979) and the WMO White Paper on impact of CO_2 on climate variations (1981).

Future studies will involve not only investigations of the deep flows themselves. Their driving mechanisms are at the surface, and they must be understood, too. High on the priority list are high-latitude deep convection in the oceans, ice formation, and shallow-shelf runoff processes, all of which occur only sporadically in time and space, and with many required preconditions on the state of the ocean and the atmosphere (see, for example, Clarke and Gascard, 1983a, 1983b.

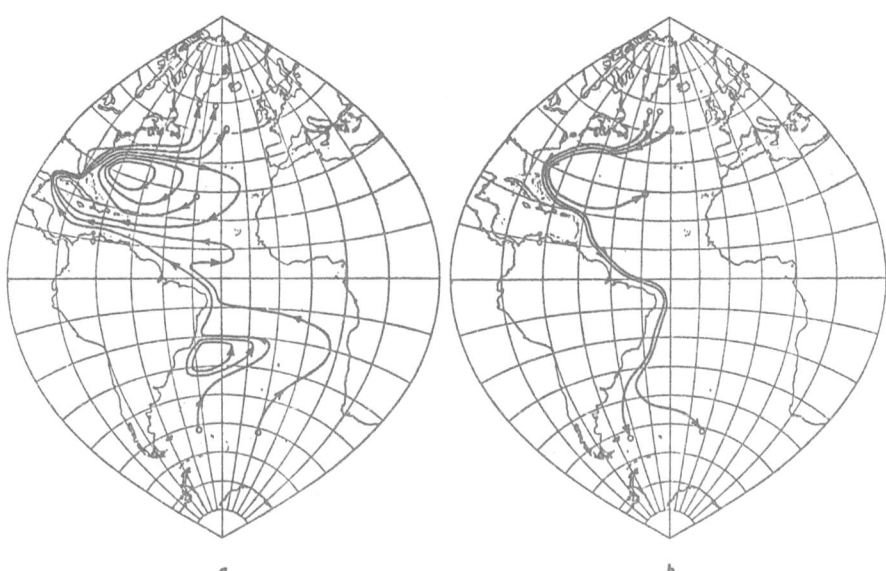

Figure 7. Schematic charts of the Atlantic Ocean, showing idealized upper ocean (a, at left) and lower ocean (b, at right), transport streamlines. Devised by Stommel (1960).

Variations in the rate at which the deep waters are formed by these processes must surely affect the deep circulation, but no one knows by how much. The subantarctic fronts have a role to play, as do low-latitude and equatorial upwelling and distributed diffusion processes (see the following section). Once again, little is known quantitatively about the dynamics of the oceanic response to forcing via such mechanisms.

Diffusion and Mixing

There are many processes by which ocean water can be mixed and thereby merge its original properties (T,s, momentum, vorticity, O_2, CO_2 and nutrient concentrations, etc.) with those of nearby water. The most straightforward mixing process is molecular diffusion, in which a water property is exchanged at a rate proportional to the spatial rate of change of the property. The resulting, so-called "Fickian", diffusion equation (see e.g. Batchelor, 1967) is

$$\partial C/\partial t = K_D[\partial^2 C/\partial x^2 + \partial^2 C/\partial y^2 + \partial^2 C/\partial z^2] \qquad (11)$$

where K_D is the "diffusion coefficient" of the water properties with concentration C (K_D has units of length^2 time^{-1}). For

molecular diffusion, K_D depends only on the property diffusing
(e.g. salt); it is not time or space-dependent.

Because the diffusivity of heat (or temperature, T) and salt
(or salinity, s) differ in the ocean ($K_{DT} \approx 100 \times K_{Ds}$),
fluids that are stably stratified in the sense that their density
increases continuously with depth, may become unstable and mix.
Two examples of such "double diffusive" processes, referred to as
"salt fingering" and "cabbeling" are thought to be of some
importance in the ocean.

Salt fingering occurs when hot, salty water lies above cold,
fresher water (see Figure 8). As soon as the interface is perturbed, the fluid which bulges down loses heat to its surroundings faster than it loses salt, becomes more dense, and sinks;
the upward bulges gain heat and rise, and "fingers" of the two
water types interleave vertically. The process is thought to
occur on a widespread scale in the ocean, where the bottom waters
(formed at high latitudes by deep convection and overflow from
polar basins) are typically fresher than the "intermediate"
waters immediately above them. Salt fingering is also considered
to be an important mixing mechanism in fronts (e.g. see Bowman
and Esaias, 1978).

Cabbeling results because the relation between the density
of seawater and its s and T is nonlinear, particularly at low
temperatures. Thus if adjacent water types mix, the mixture may
be denser than either, resulting in further mixing by convective
overturning and turbulence as the mixture sinks. Such processes
are important in the formation of Antarctic Bottom water. For a
recent review of double diffusion see the article by S. Turner in
Warren and Wunsch (1981). For more details, see Turner's (1977)
book.

Turbulent Mixing

Turbulence is by far the most efficient mixing process in
the oceans. When a fluid becomes turbulent, it is continually
being deformed by random fluid motions. These deformations-
-spinning, stretching, interleaving -- cause an originally con-
tiguous parcel of water to be formed into fine sheets or fila-
ments. Thereby gradients in water properties between neighboring
parcels are continually sharpened, since the process of filament-
ation brings them into ever-closer proximity, so that molecular
diffusion can occur with great efficiency.

Turbulence can be generated in a number of ways, either by
direct mechanical "stirring", or through the action of buoyancy
forces. One source of turbulence in the ocean is the breaking of
sea surface waves. Such turbulence carries with it momentum from

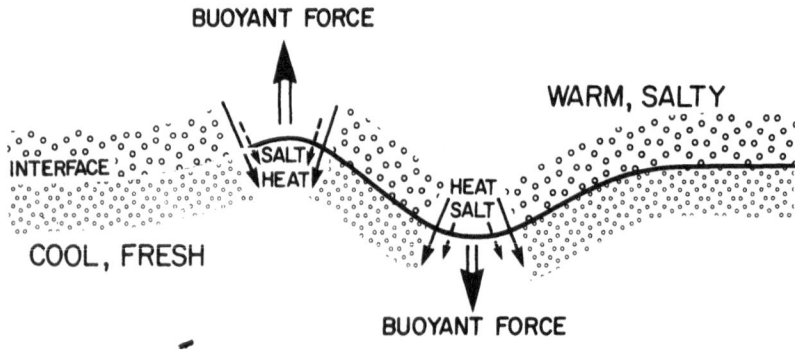

Figure 8. Salt fingering: a double-diffusive process known to
be important in the ocean, particularly in the vicinity of
fronts. Because heat diffuses 100 times faster than salt across
the (initially sharp and horizontal) interface, any vertical perturbations result in buoyant forces which further perturb the
initial state. Upward extensions of fresh (i.e., low-salinity)
water are heated and rise; downward extensions of salty water are
cooled and sink.

the wave field, and as well as mixing the surface water to a
depth of one or two wave amplitudes, it transfers the wave
momentum as a current to the upper mixed layer (see the section
on the Oceanic Mixed Layer). The action of a velocity shear can
also cause turbulence. Buoyancy forces can act to stabilize the
water column, if heating from above or sea surface precipitation
are dominant in determining the vertical variation of water
density, or to destabilize the water column, if cooling from
above or sea surface evaporation are dominant. In the presence
of an otherwise stable density gradient, a laminar flow will
become unstable when it reaches a critical ratio of buoyancy-
induced stability, to shear-induced "mixing capability". For
vertical shears the ratio (one of many such "numbers" having to
do with fluid stability) has been named after L.F. Richardson,
and it is defined as

$$Ri = (-g \, \partial\rho/\partial z)/\rho(\partial u/\partial z)^2 \qquad (12)$$

This particular version is called the "gradient" Richardson
number, since it is defined in terms of gradients. If the shear
is large ($\partial u/\partial z \gg 1$), then Ri becomes smaller, and hence a fluid
is more likely to be turbulent if the Richardson number is small.
Large density gradients ($-\partial\rho/\partial z \gg 1$, i.e. increasing density
with depth) cause Ri to be larger, and hence turbulence is
suppressed for large Richardson number.

Whenever density gradients exist in a fluid, internal waves can exist on the gradient (precisely as sea surface waves exist on the enormous air-water density gradient). They oscillate with displacements of the isopycnals from their mean depth with a frequency

$$N = \left\{-(g/\rho)\ (\partial \rho /\partial z)\right\}^{1/2} \tag{13}$$

N is called the Brunt-Väisälä frequency, and

$$Ri = N^2/(\partial u/\partial z)^2 \tag{14}$$

Engineers often work with the inverse square root of Ri; it is called the "Internal Froude Number" and is usually designated Fr.

The criterion for stability is not simply determined, either theoretically or experimentally, in the laboratory or for field work: there are many types of shear-buoyancy instabilities. In general, flows are stable for Ri > 1 and can be unstable for Ri $\leq$ 1/4.

MESOSCALE OCEANOGRAPHY
Ocean Eddies

Eddies in the ocean are the dynamical equivalent of "synoptic-scale" disturbances, or storms, in the atmosphere. They differ in important ways from their atmospheric analogs. There are no "fronts" in the interior of the eddies, as there are in typical high-latitude cyclones; in the ocean the frontal processes are confined to the regions of large current shear at the edges of the eddies. In the ocean, there is no analog to the evaporative/condensative processes which play such a central role in the dynamics of atmospheric storms. Oceanic eddies, once formed (often by instabilities in the western boundary currents), simply rotate, slowly converting their initial store of potential energy to kinetic energy. The scale of such disturbances has been defined by Rossby (1938) to be the distance over which a disturbance will be directly transmitted by pressure gradients, before the effects of the earth's rotation become appreciable. It is defined as the ratio of the speed of the wave in the medium to the Coriolis parameter, which for a stratified fluid is

$$L_R \equiv \text{Rossby deformation radius} = (gh\Delta\rho/\rho)^{1/2}/f \tag{15}$$

where h is the vertical extent of the disturbance and $\Delta\rho/\rho$ is the degree of density stratification. L_R is about 800 km for the atmosphere and 50 km for the ocean.

Oceanic eddies have been a central concern of oceanographers over the 1970-1980 decade (see e.g. Oceanus, 1976). The MODE and POLYMODE experiments were designed to elucidate the role of meso-scale processes in ocean dynamics, and have greatly increased our understanding of such processes. Ocean dynamic models are now in existence (Holland and Rhines, 1980) that resolve the oceanic eddies and incorporate their dynamics into the general oceanic circulation. Such models, although they cannot yet be said to be predictive, do reproduce the general circulation remarkably well.

Gulf Stream Rings

Figure 9a describes the birth of a cold-core ring from the Gulf Stream; figure 9b shows a typical cross-section for a cold-core ring. Similar (but not identical) rings are spawned by the Kuroshio system (White and McCreary, 1976). The rings form as the result of instabilities in the strong currents, which begin as "meanders" and end as pinched-off "rings", which can be either cold-core and rotating counterclockwise (in the Northern Hemisphere) when pinched off equatorward of the stream, or warmcore and rotating clockwise when pinched off to poleward. Both types have now been extensively observed, most effectively with satellites (Figure 10) and with drifters that report their positions regularly to a satellite (Richardson et al., 1977).

The dynamics of such rings are understood in general, but their interactions with the general circulation, and hence with the global climate system, is not well-understood at all. A cold-core ring, for example, consists of a well-defined area of water 50 to 100 km in diameter, typically cooler by about 5°C than its surroundings, rotating with tangential velocities up to 1 ms^{-1} in the core. In the Gulf Stream the typical vertical extent of the rings is from the surface to 1500 m. The density distribution within the rings shows a general raising of the main pycnocline by as much as 500-600 m, and a sea surface depression of 0.4 to 0.5 m. Both contribute to a geostrophically balanced counterclockwise flow in the Northern Hemisphere. Warm-core eddies rotate clockwise in the Northern Hemisphere, and are characterized by a sea surface elevation and depression of the main thermocline, both of the same magnitude as for the cold-core eddies.

The eddies are spawned from the Gulf Stream at the rate of 5 to 8 per year on each side of the Stream. Their life time, before decaying or being reabsorbed, is up to two years for cold-core, and less than one year for warm-core rings. They move at rates of 2 to 10 km/day, and, in the generally southerly Gulf Stream recirculation region tend to move in a southwesterly direction.

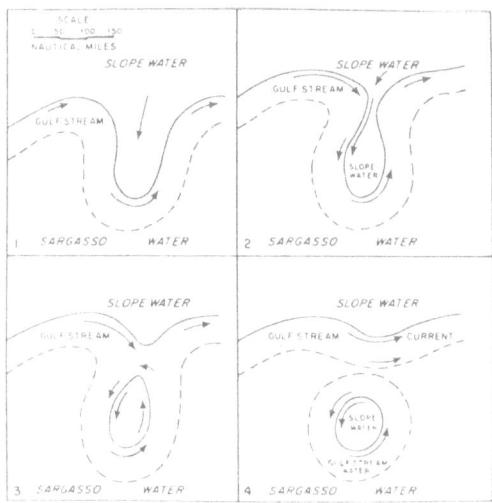

Figure 9a.   Schematic of the birth of a cold core ring from the
Gulf Stream.   (After Parker, 1971.)

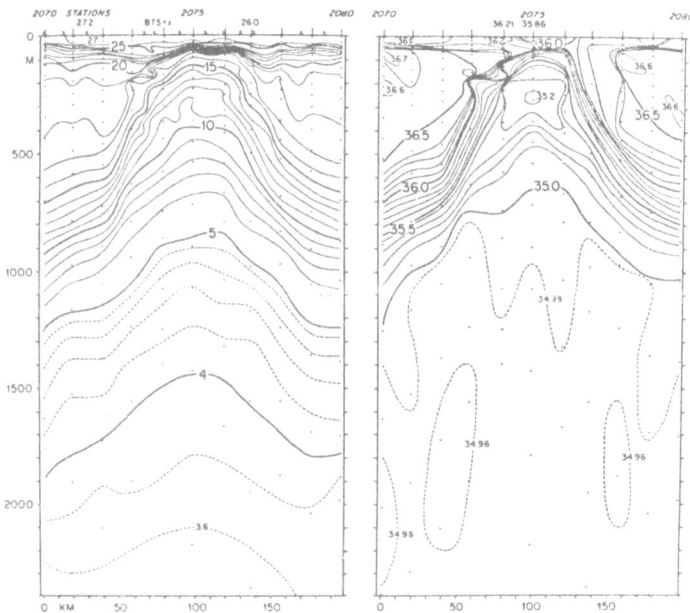

Figure 9b.   Temperature and salinity sections through a Gulf
Stream "cold core" ring, or eddy.   Location is 36.5°N, 64°W, on
31 July 1967; data from R.V. "Crawford" cruise 158.   (After
Fuglister, in Angel, 1977.)

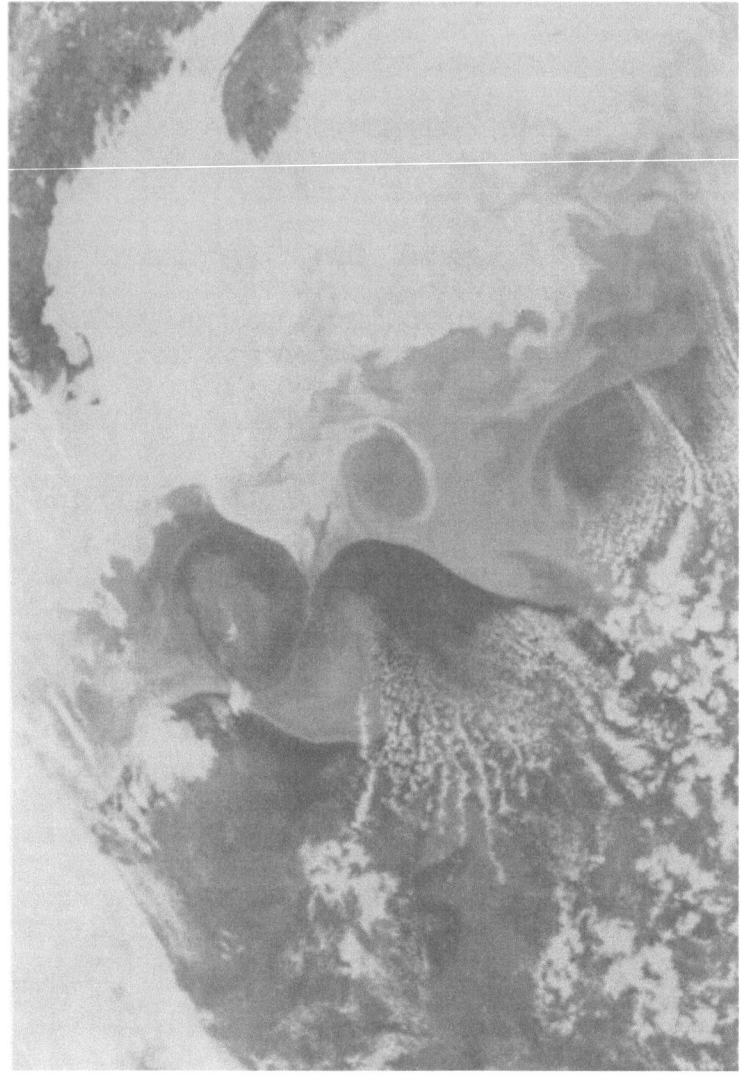

Figure 10. NOAA-5 infrared image of the Gulf Stream off Cape
Cod, MA. The lighter-coloured waters are cool; darker-coloured
regions are warm, Gulf Stream water (the white areas in the lower
right are clouds). Several warm-core eddies are evident; the one
at the center is particularly well-formed. (Photo courtesy of
Atlantic Environment Group, U.S. National Fisheries Service,
Narragansett, RI).

To see the reason for the southwesterly drift, consider a warm-core, counterclockwise-rotating eddy which is in geostrophic equilibrium, so its pressure gradients are balanced exactly by the mean Coriolis force at its central latitude. Since water equatorwards of the central latitude has higher counterclockwise angular momentum than the water polewards of the central latitude, the equatorwards water will turn poleward with a slightly smaller radius of curvature than that of the polewards water. This effect, due to the latitude variation of the Coriolis parameter, causes all eddies, warm-core and cold-core, to propagate westwards.

Rings cause significant disturbances in their environment, particularly when they approach the shore, as warm-core Gulf Stream rings do. They interfere with frontal and long-wave generation processes at the shelf break, and with surface layer circulation and mixing, and hence with mass exchange over the continental slope and shelf (Csanady, 1979). Dissipation of such rings and the lateral mixing they cause may be an effective way of transporting heat, salt and angular momentum in the ocean. Eddies decay by using up their store of potential energy by converting it to kinetic energy of spinning (in a young cyclonic ring the main thermocline is raised 500-600 m, giving it an "available potential energy" of 10^{17} joules; using this up in 2 years, averaged over a 50 km diameter eddy, gives an energy dissipation rate of 0.8 W m^{-2}). Eddies are known to be an integral part of ocean dynamics, but their overall significance is not yet fully understood.

Eddy motions are important for another reason: they make sampling very difficult. N. Fofonoff, in Warren and Wunsch (1981), states "The Gulf Stream, if it exists near the bottom above 70°W ..... is nearly completely masked by the strong deep eddy field". P. Rhines, in Oceanus (1976) writes, referring to the entire ocean, "The attempt at understanding the balance of forces and flow of energy in the ocean is made difficult by the variability of the currents: that is, by the eddies. The currents are in fact far too capricious to be mapped once and for all, so that the procedure of the classical oceanographer - the gradual filling in of a jigsaw picture of the ocean - may simply not work". He then goes on to suggest that overcoming such a fundamental difficulty will involve the formulation of theories that describe the eddy fields statistically rather than causally. Meanwhile, existing eddy-resolving models (e.g. Holland, 1978) strain the largest available computers to their limits, and even remote sensing techniques such as satellite altimetry (Wunsch, 1981) have a difficult time resolving eddy-scale motions in time and space except in a statistical sense.

The eddies have been found to be ubiquitous, at least in the Northern oceans, and oceanographers, fully aware of the sampling problems eddies create, are presently reassessing their "error bars" and their experimental strategy. More than any other single phenomenon, eddies have changed the face of theoretical and experimental oceanography. The new age belongs to the statistically-oriented numerical simulation, the drifting float, and the satellite. Deep soundings from ships, although far from outliving their usefulness, must now be interpreted from an entirely new perspective: is the resulting field really a time (or space) average, or is it the result of sampling a field of eddies?

Frontal Processes

Oceanic fronts are best defined as boundaries between water masses with dissimilar properties. The fronts are often marked on the surface by disturbances of the waves there or by lines of foam or debris. They generally have associated with them regions of convergence and relatively strong vertical motions. Such circulations, as well as the various diffusive processes and the large gradients of water properties at such boundaries, form effective mixing agents. Consequently, fronts are considered to be major contributors to the formation of new water masses.

Estimates of the volumes of water mixed within the various types of fronts are, unfortunately, not yet available. The small-scale processes by which mixing occurs within fronts are the subject of intense study (for an excellent recent review see Garrett, 1982), but only crude estimates, such as that given for the decay rate of eddies in the section on Gulf Stream rings, are available now. Fronts occur not only at the sea surface, but also at any depth in the water column, including the vicinity of the sea bottom. They occur on all spatial scales, from less than a meter to thousands of kilometers. In general they are caused and maintained in the oceans by processes similar to those in the atmosphere, but fronts in the ocean take on a greater diversity of form and strength than they do in the atmosphere.

Following Bowman and Esaias (1978), fronts may be classified into six types, according to their scale and dynamical origin.

Planetary fronts are those associated with large-scale convergence of surface Ekman transports, and are typically found in mid-ocean. A good example is the Antarctic convergence zone, or Polar Front (see Figure 11a and b), which is at about 60°S and forms the northern limit of the so-called Antarctic Surface Water formed by ice-melt in summer and surface cooling in winter.

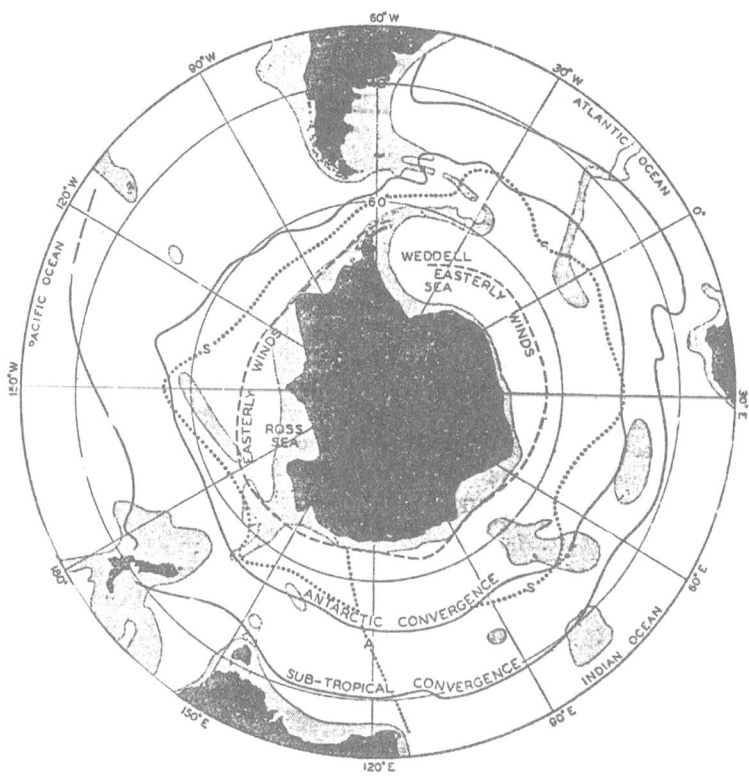

Figure 11a. Map of Antarctica, showing (solid lines) the position of the Sub-tropical and (winter) Antarctic Convergence
Zones, and (dotted line) the summer Antarctic Convergence zone.
the dashed line shows the outer limits of the Antarctic easter-
lies. The roughly meridional dotted line marked "A" is the position of the line of stations from which Figure 11b is derived.

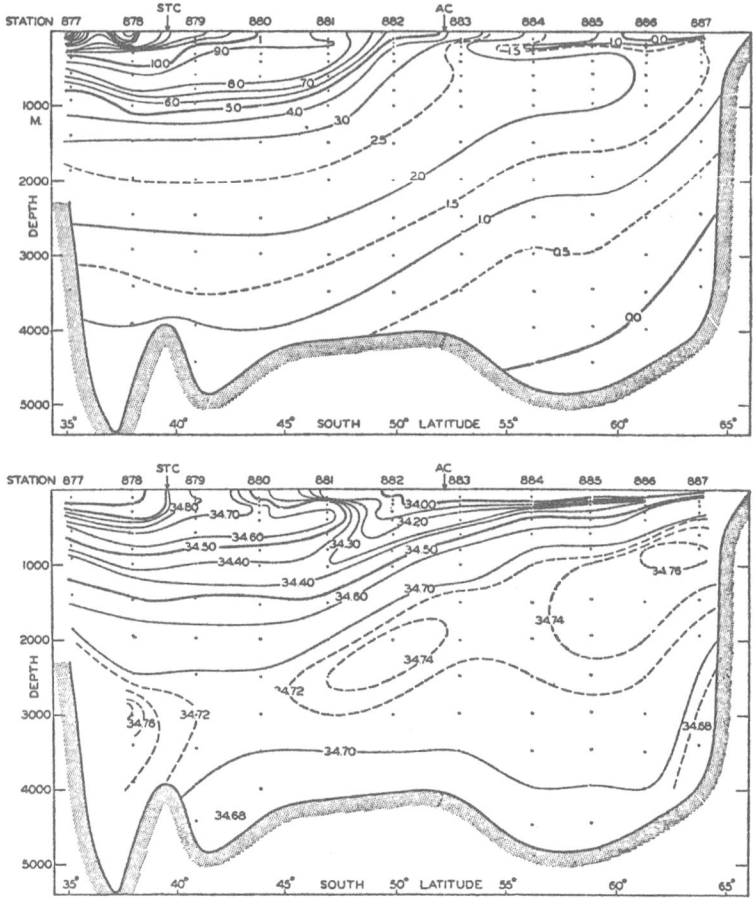

Figure 11b. Temperature (top) and salinity (bottom) sections
from Australia (on the left) to Antarctica (section marked "A" on
Figure 11a). Positions of Sub-tropical and Antarctic Convergences marked "STC" and "AC". (After Sverdrup, Johnson and
Fleming, 1942.)

Western boundary current edges often behave as fronts.
Warm, salty tropical water is brought into relatively close
proximity with cold, fresh high-latitude water by the western
boundary currents. Sharp boundaries and mixing on all but the
planetary scales is the inevitable result. For a beautiful
example of such a front, see the frontespiece to Stommel's (1960)
book; see also Figure 10. That such fronts are important contributors to oceanic dynamics and mixing is discussed further in
the preceding sections on "Ocean Eddies" and "Gulf Stream Rings".

Shelf break fronts are formed at the edges of continental
shelves, where waters characteristic of the shallow shelf regions
meet and mingle with water types typical of the continental
slopes and the deep ocean. All types of diffusion and mixing
occur, and the flows can be either baroclinic (surfaces of constant pressure and density tilted) or barotropic (isobars and
isopycnals parallel) depending on whether or not the salinity and
temperature fronts coincide, and which controls the water
density.

Upwelling fronts (see Figure 12) are generally associated
with offshore Ekman transports generated by alongshore winds.
They are characterized by a pycnocline that reaches the surface
with surface water offshore and deep water at the surface inshore. Such fronts commonly occur off the West coasts of continents (California, Peru, Africa). They interact strongly with the
local meteorology, and are one of the few locations on the globe
where waters beneath the permanent pycnocline come into direct
contact with the atmosphere. They thus act as an important path-
way by which airborne gases and particles can be exchanged with
the deep ocean.

Plume fronts occur on the boundaries of the fresh-water dis-
charge plumes produced by large rivers, such as the Amazon and
Columbia Rivers. They act to stir the fresh runoff waters, with
their load of terrestrial solutes, into the ocean.

Shallow-sea fronts are formed in shallow seas and estuaries
and around islands, shoals, capes, etc. They form the boundary
between well-stratified offshore waters and waters in the shallow
seas, which are well-mixed by winds and tides. Simpson and
Hunter (1974) have defined a useful "stratification ratio", R, as
the rate of production of potential energy by surface heat flux Q
to the rate of tidal energy dissipation:

$$R = (gh\beta_T \, Q/2c_p\rho)/(C_d u^3) \approx h/u^3 \tag{16}$$

providing Q is constant. Here β_T is the volume expansion
coefficient of the water, h its depth, c_p its specific heat at

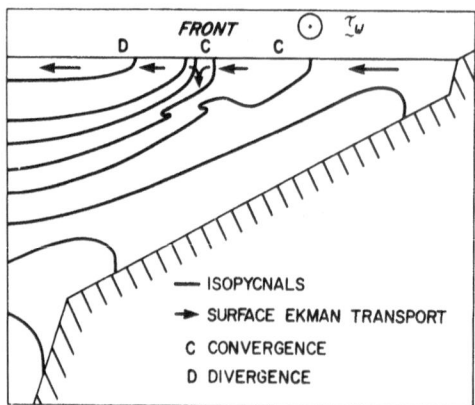

Figure 12. Diagram of an "upwelling front" in the Northern
Hemisphere. A wind stress τ_w, out of the paper, drives surface
water offshore ("surface Ekman transport" arrows). The dynamics
by which the front is maintained between the inshore upwelled
deep water and the offshore surface water causes convergence and
sinking at the front ("C") and divergence offshore ("D"). (After
Bowman and Esaias, 1978.)

constant pressure, ρ its density, and C_d a "drag coefficient"
relating the drag exerted by the bottom on the water above it to
the square of the water speed u. (The reader should note that R
is really a type of Richardson Number: see the section on
Turbulent Mixing, equation 12.) Plots of the logarithm of h/u^3
are commonly used to distinguish well-stratified (large R) from
well-mixed (small R) areas, and the fronts that lie between them.
Figure 13, from Bowman and Esaias (1978) is an example of such a
"h/u^3" plot; fronts are visible on the eastern sides of the
Celtic and Irish Seas, and are also present in the nearshore
areas on the south side of the English Channel.

Tides and Large-Scale Waves

Three excellent reviews of tidal motions are "Ebb and Flow"
by Defant (1958), "The Tides" by Darwin (1968 edition), and the
more technical review by Cartwright (1977). The forces which
generate the tides can be described from the dynamics of the
earth-moon system.

The earth and the moon are two spheres rotating around a
centre of mass that is within the earth but is not at the earth's
centre, being shifted towards the moon. The earth itself (which
also spins, a fact that we ignore for the moment) then rotates <u>as
a solid body</u> (like one end of a "dumbell") about the centre of
mass of the earth-moon system, and each point on the earth's
surface experiences a centrifugal force inversely proportional to

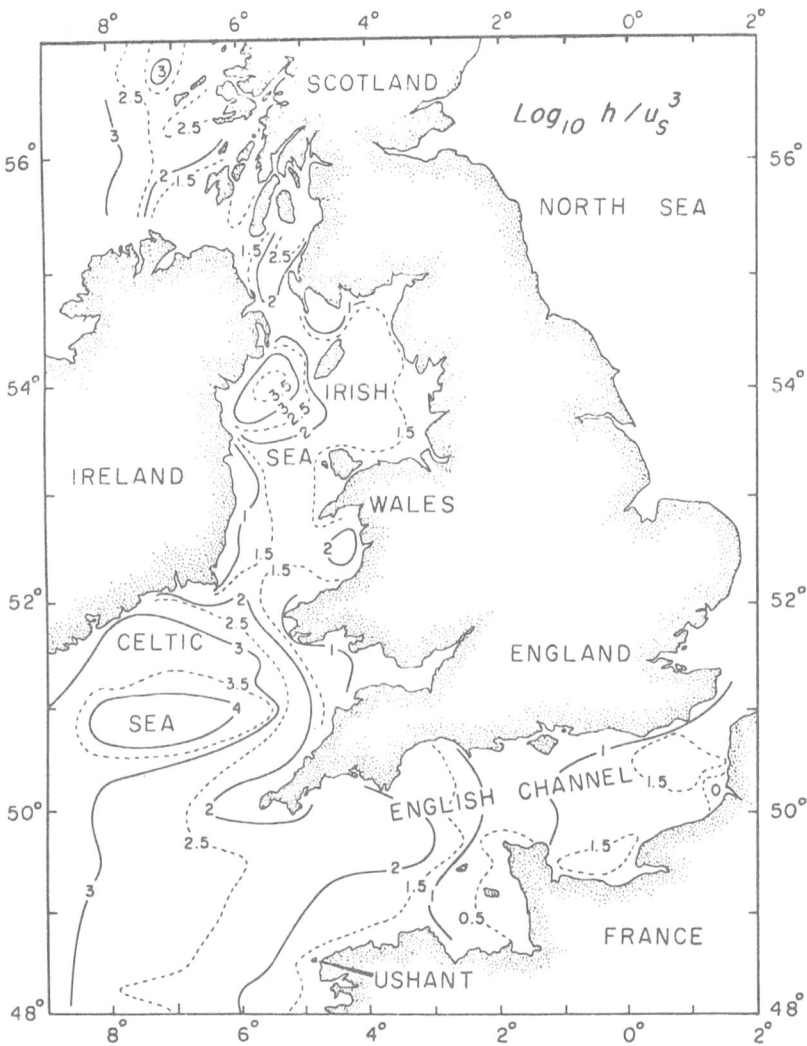

Figure 13. Plot of the $\log_{10}$ of the tidal mixing parameter h/u_s^3, where h is the water depth and u_s is the surface tidal speed of the water, for the seas of southwest England and Ireland. Tidal mixing fronts are found between regions of little mixing (Celtic Sea, western Irish Sea) and strong mixing the coast of South Wales, eastern Irish Sea), and in the vicinity of major headlands (Land's End and Ouessant or "Ushant"). (After Bowman and Esaias, 1978.)

its distance from the centre of mass and directed away from the
moon. Each point experiences the moon's gravitational pull as
well as the earth's. The only asymmetrical forces are the moon's
pull (the side of the earth away from the moon is about sixty-one
earth radii from the moon, and that facing the moon if fifty-
nine, so the gravitational pull is stronger on the side nearest
the moon) and the centrifugal force caused by the earth's solid-
body rotation (since the centre of mass of the earth-moon system
is shifted toward the moon from the earth's centre, the centrifugal force is stronger on the side of the earth furthest from
the moon). When the tidal forces are added, there is a net out-
ward force on the earth's surface at the points nearest and most
distant from the moon, and net inward forces on points on the
earth at right angles to them. As the earth spins, these forces
act on the oceans, causing two tides per day. The tidal forces
are not just those from the moon but also from the sun (the sun's
force is about half that of the moon, because it is so much
further away from the earth).

The tidal forces have various distinct frequencies, caused
by perturbations in the orbits of the earth-moon-sun system. The
actual magnitude of the tide at a given location is determined by
interactions of the tide-generated waves with local topography.
Since they are "body" forces, the tidal forces act everywhere,
not just at the sea surface; there are measurable tides in the
earth itself, as well as in the atmosphere and on pycnoclines in
the ocean. The latter are called "internal tides", and are
clearly evident in most oceanic current meter records. They are
generated by direct forcing and also indirectly, as the sea surface tide (the "barotropic" tide) impinges on continental
shelves, causing the pycnocline to oscillate and radiate "baroclinic" tidal-frequency waves seaward.

The wavelength of tidally-generated waves is half the circumference of the globe; since the depth of the ocean is only a
small fraction of a wavelength, they obey the equations for
shallow-water waves. There are various other waves, also
shallow-water by nature, which occur in the ocean: "Poincaré"
waves, coastally-trapped "Kelvin" waves, and others. Their
exposition lies beyond the scope of this text, and the interested
reader is referred to texts on physical oceanography or geophys-
ical fluid dynamics (e.g. LeBlond and Mysak, 1978 or Gill, 1982).

Oceanic tides are important in transport only in the sense
that they induce mixing and generate internal waves, which in
turn break down to produce mixing. Most tidally-induced mixing
occurs at the edge of or on the continental shelves. The so-
called "h/u " criterion for determining where tidal mixing is
likely to be important is discussed in the section on Frontal
Processes. Such mixing in shallow seas can form new water

masses, particularly at high latitudes, where warm, salty waters delivered from low latitudes by western boundary currents can be mixed tidally with cooler, fresher northern waters. The eastern continental shelves of North America are breeding grounds for several at least partly tidally-generated water types, as are the shallow waters of the Barents and East Siberian Seas in the Arctic.

Internal Waves

Gravity waves occur on all density gradients in the atmosphere and the ocean; in the latter the "internal waves" on various density gradients are as ubiquitous as are surface gravity waves. They have wavelengths from hundreds of meters to tens of kilometers and periods from tens of minutes to tens of hours. Their amplitudes can be tens of meters. The currents and current shears associated with them often interact with short-wavelength (≈ 10 cm) surface waves, producing surface patterns of alternately ruffled and unruffled water that are highly visible to the (elevated) eye or to obliquely incident radars (see e.g. Fu and Holt, 1982). The large displacements of isopycnals (and hence isohalines, isotherms, and other isolines) are an extremely important source of sampling error to oceanographers. Classical geostrophic transports are calculated from salinity and temperature information collected from specific depths at various locations. It is easy to see how an internal wave, by displacing isolines vertically by distances of 10 m or more over distances of one or more km, could change the outcome of such calculations, particularly since internal waves can propagate with speeds of 0.1 m s^{-1} or more.

Classical internal waves (that is, those occurring on a relatively sharp density discontinuity) have their largest vertical displacements, and hence their largest horizontal velocity differences, or shears, at the discontinuity. Such velocity shears are known to become unstable (see the section on Mixed Layers) when the ratio of the stabilizing influence of the density difference to the destabilizing influence of the shear (i.e. the Richardson Number Ri) becomes small. Below Ri $\approx 1/4$, internal waves begin to break and to mix the waters above and below the interface. This mixing mechanism is important in the ocean, since internal waves and shears are so common. Internal waves have been extensively studied in the ocean and a number of useful (although often somewhat mathematical) reviews are available. Phillips (1977) gives a relatively complete and authoratative coverage. Perhaps the most useful reviews for estimating quantitatively the effects of internal-wave mixing in the ocean are those of Munk (1966) and Garrett (1979).

THE OCEANIC MIXED LAYER
Introduction

The oceanic mixed layer is defined as the region of the
upper ocean which is influenced by mixing processes, with predominant seasonal and daily cycles, which originate at the sea
surface. It has uniform properties and is bounded below by a
large density gradient, called the "pycnocline" (gradients in
temperature are "thermoclines" and gradients in salinity are
"haloclines". For dynamical reasons discussed further in the
section on Thermohaline Circulation, the ocean, the density of
which is controlled primarily by temperature, has a "permanent
thermocline", delineating the lower limit of the annual cooling
cycle, at depths ranging from less than 100 m in some equatorial
regions to more than 1000 m in the centre of the large oceanic
gyres. Above this a series of seasonal thermoclines are formed,
which have the permanent thermocline as their limiting winter
depth, and are reformed at the sea surface every spring. The
various density discontinuities of the mixed layer, and in particular the permanent thermocline, act as barriers to the downwards mixing of substances in the mixed layer because they are
very efficient at damping out buoyant convection. The "rate
constants" of the various pycnoclines for various mixing processes can at the moment only be guessed at, or inferred from
large-scale distributions of the various tracers.

The atmosphere, like the ocean, has a mixed layer, but the
dynamics of the phenomena in the two media are dominated by
different processes. In the atmosphere, mechanically-generated
turbulence has its source mostly near the sea surface, from
instabilities in the strongly sheared flow there. In the ocean,
the waves generate some turbulence, but equally important sources
are convection and shear instabilities at the base of the mixed
layer. In the atmosphere, the clouds have a strong influence on
absorption and transmission of solar energy, both incoming short-
wave and outgoing longwave. In the ocean, most of the incoming
solar radiation is absorbed within one meter of the surface, and
infrared absorption and emission is in the top millimeter. Solar
heating causes instability in the air and stability in the water;
evaporation causes convective instability in both media.

The oceanic mixed layer is of intense interest because its
upper surface is the region of first interaction of the various
gases and particles with the ocean. Heat and salt are stored;
CO_2 is chemically altered and stored; the various atmospheric
daters and tracers start to be mixed into the ocean here. The
time scales are seconds to years, the vertical scales are centi-
meters to decimeters, and these scales overlap with those of the
surface (on the short end) and the deep ocean (on the long end).

Dynamically the mixed layer is a terra incognita. Velocity measurements (or good density measurements) are very difficult to make there. The principal measurement difficulty, both in the fluid and as an influence on measurement platforms, results from surface wave motions. The mixed layer is a region of strong air-sea interactions, and hence is governed by strongly nonlinear dynamics. (Surface and internal wave breaking, mixing of heat and salt across isentropic surfaces ("diabatic" mixing), particularly in coastal regions, advection, absorption and re-emission of solar radiation, and strong biological activity all contribute to the dynamics). This makes analytical solutions unlikely. Consequently, the approach has been to parameterize the strong effects in terms of measurable quantities and do simple budgets.

Time and Space Scales

Figure 14, from Tabata and Giovando (1963), shows the time variation of the depth of the oceanic mixed layer at Ocean Weather Station "P" (50°N, 145°W, in the North Pacific) averaged over four years. Note that, in contrast to the atmospheric mixed layer, the oceanic mixed layer is deeper in winter than in summer. Figure 15, after Woods, 1983, shows schematically the yearly cycle of heating/cooling, including the range of diurnal variation. There are also large spatial variations in the observed depth of the mixed layer, as indicated in the Robinson, Bauer and Schroeder (1979) atlas (Fig. 16a).

Since the convectively mixed layer contains, distributed within it, all gases and particles transferred from the atmosphere to the ocean over a time interval short compared with the equilibration time of the deep ocean, it will pay to understand the dynamics of the mixed layer well.

Modelling the Mixed Layer

The modelling of mixed layers is actively evolving, after a big push in the mid-70's by a whole series of "one-dimensional" models, which are detailed in Kraus (1977). These one-dimensional models use as their basic premise the observed fact that vertical gradients are much stronger than horizontal ones at most places in the ocean. Hence the mixed layer can be thought to have a vertical but no horizontal structure, and the forces at work can be considerably simplified. Time variation is, of course, allowed. Models allowing for both time and three-dimensional space variation are being developed and tested now. They must allow for advection: the transport of water and its properties from place to place by oceanic currents, either driven by large-scale forcing external to theproblem (Ekman flux divergence, for example) or by local forcing due to wind and/or

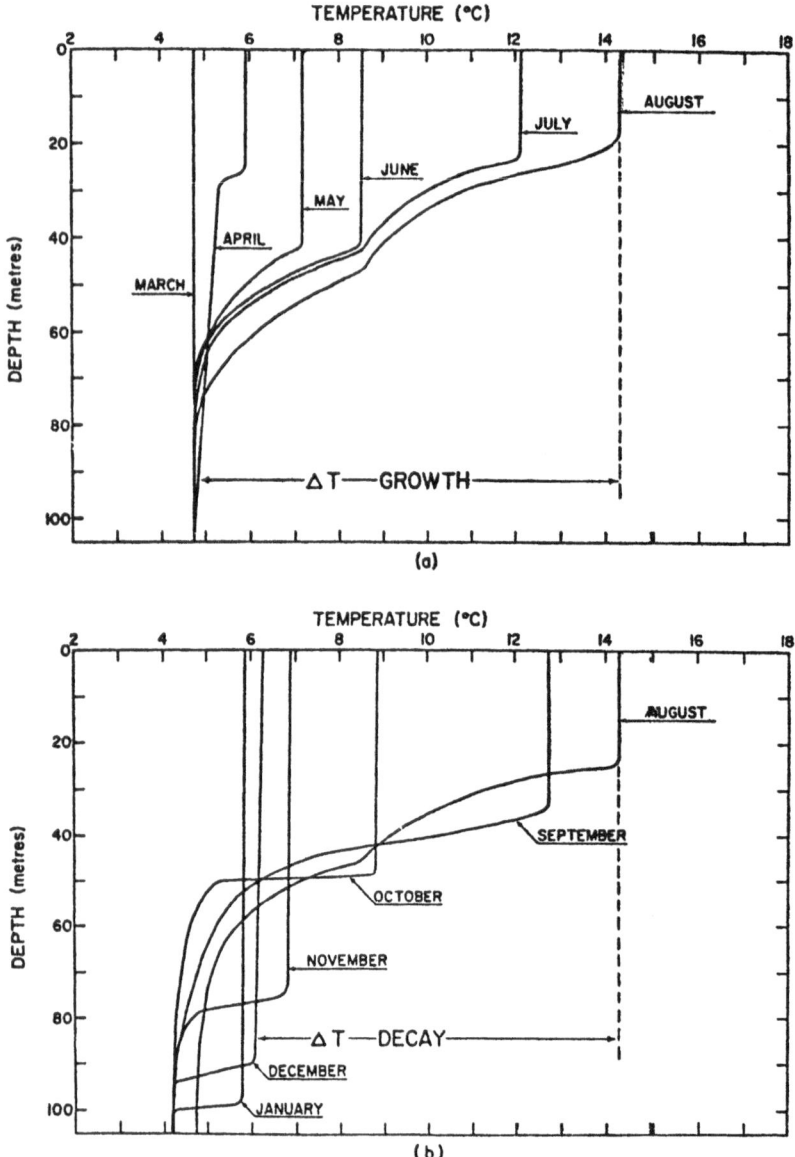

Figure 14. Time variation of mixed layer thermal structure at
Ocean Weather Station "P". (a) March-August, (b) August-
January. (After Tabata and Giovando, 1963.)

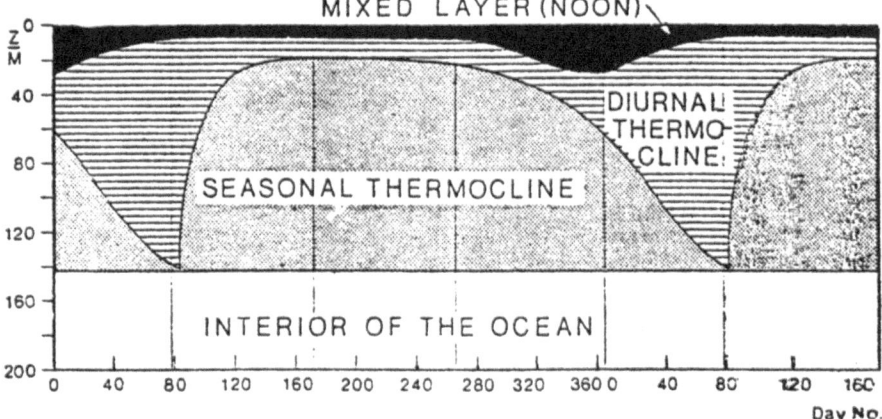

Figure 15. Idealized yearly cycle of mixed layer depth, including depth variations of the noon thermocline (dark), the diurnal thermocline (hatched), and the seasonal thermocline (stippled). (After Woods, 1983.)

buoyancy effects. An illustration of the results of one of the earlier models (Zubov, N.N. in Gorshkov, 1978) is given in Figure 16b. Comparison with observations (Fig. 16a) shows that the models are far from perfect.

Forcing Variables

The mixed layer is forced primarily by solar heating by shortwave radiation, the wind stress, and the vertical fluxes of latent and sensible heat through the sea surface. Smaller but in some circumstances important driving forces are precipitation and the mixing effects of internal waves. The presence of slicks affects both the heat fluxes and the influence of surface waves. The response of the mixed layer to such forcing is radiation back to the atmosphere at long wavelengths, downwards turbulent mixing, including entrainment at the bottom of the mixed layer by mechanical and buoyant forces, and the generation of currents on scales ranging from oceanic (Ekman flux divergence) to microscale (turbulence).

Solar Radiation

Solar heating is a complex process. Once the spectrum of radiation reaching the sea surface after transmission through the atmosphere is known, and once the reflectivity (or albedo) of the sea surface is known, then an attempt can be made to estimate the scattering and the penetration depths (and thus the heating capabilities) of the various wavelengths of the incoming radiation. The necessary procedures, and the difficulties involved, are succinctly summarized in the article by A. Ivanoff in Kraus (1977), and given in more detail in the books by Budyko (1974) and Jerlov (1976). Figure 17, from Augstein, 1981, shows the

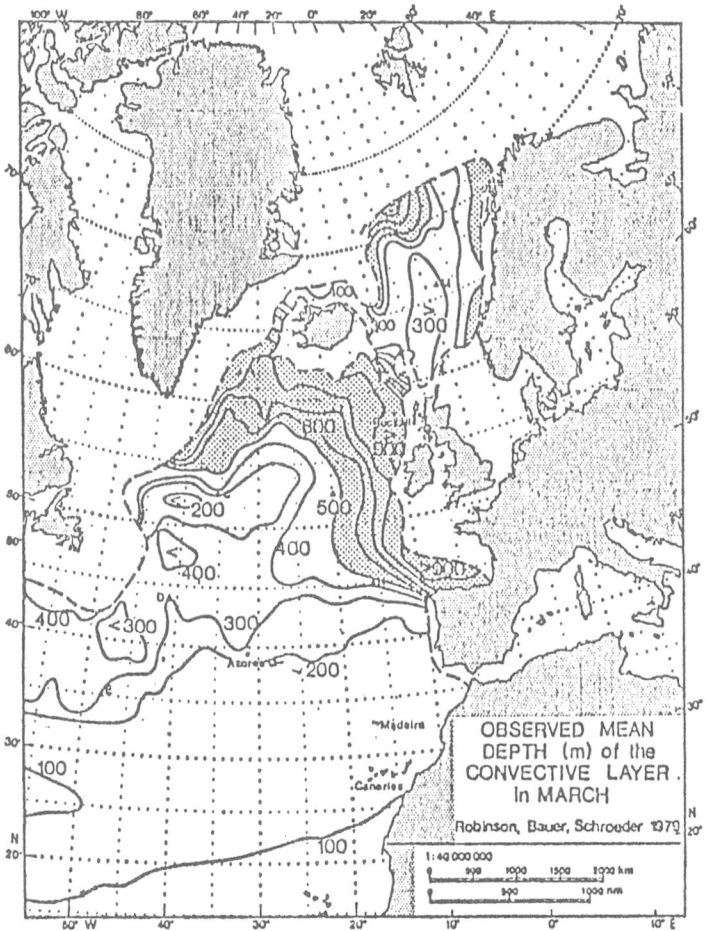

Figure 16a. Variation of the annual mean depth reached by the
mixed layer for the Northeast Atlantic Ocean. The region of
total ice cover is crosshatched. (From Woods, 1983; after
Robinson et al., 1979.)

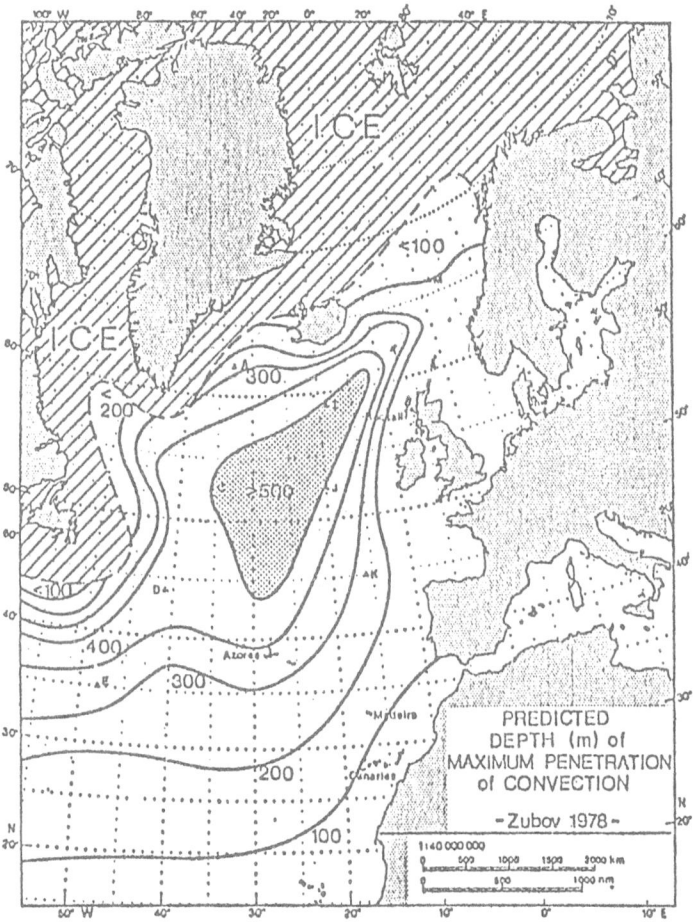

Figure 16b. Prediction for the annual mean depth reached by the
mixed layer for the Northeast Atlantic Ocean, using a mixed-layer
model attributed to N.N. Zubov. (From Woods, 1983; after
Gorshkov, 1978, Plate 141.)

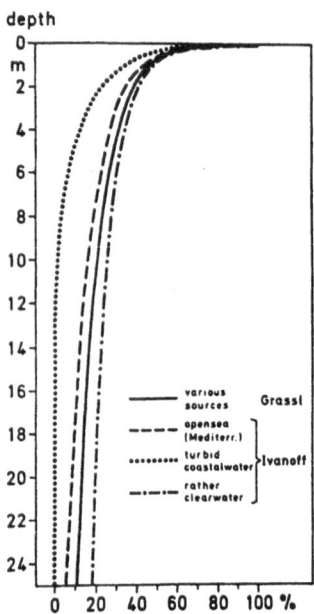

Figure 17. Percentage of incident solar energy (allowing for tha
reflected) absorbed per meter of water as a function of depth.
The solid curve is from H. Grassl (pers. comm.); the other curves
are from the article by A. Ivanoff in Kraus, 1977. (After
Augstein, 1981.)

fraction of incident solar energy (allowing for that reflected)
absorbed per meter of water as a function of depth, for different
types of seawater. It is plain to see that a large fraction of
the incident solar energy is absorbed in the first meter. Also
the "turbidity" of the water, which is strongly influenced by
biological and nearshore processes, is important in determining
the vertical distribution of solar heating in the mixed layer.
How to model these effects remains one of the central questions
in mixed-layer theory.

Wind Forcing

        Wind forcing introduces a number of effects. First, the
wind produces surface gravity waves, and although their effects
diminish exponentially with depth, they produce orbital currents
which typically exceed other currents by factors of 2 to 10 in
magnitude. Of no small consequence are the motions induced by
the waves on measurement platforms. The reader is referred to
articles by J. McCullough, R. Davis and R. Weller, W.
Blendermann, and H. Berteaux in Dobson, Hasse and Davis (1980)
for discussions on wave-induced errors in current meters and the
wave-induced motions of buoys and moorings.

It remains unclear how the momentum and energy of the wind is transferred to mean currents in the open ocean. One possibility (Dobson 1971) is that it is first transferred to the waves and then, through wave dissipation, "handed over" to the mixed layer below. In any case, the "hand-over" process is a complex one with many possibilities for fluid motion. The small-scale wave-generation and dissipation processes are dealt with later; we will concentrate here on the larger-scale phenomena. Pollard (1970) has investigated the generation of inertial oscillations (horizontal circular motions of the entire mixed layer, which result whenever the water is forced to move impulsively, followed by a removal of the forcing) in the ocean. He finds that the energy of the oscillations is confined almost entirely to the surface mixed layer (the oscillations, as we shall see, help to create the mixed layer); the amplitude of the oscillations is strongly affected by the initial depth of the mixed layer; it is nearly independent of the horizontal scale of the wind field or of the vertical stratification in the mixed layer. What matters is the time history of the wind stress vector. If the wind speed and direction vary on time scales much less than one inertial period (f^{-1} sec; about 17 h at 45° latitude), then inertial oscillations result. If, for instance, a cold front embedded in a low pressure area passes a given point in the ocean, then the sharp clockwise (in the Northern Hemisphere) shift in wind direction associated with the front will efficiently generate clockwise inertial oscillations, which will then persist until they disperse out of the region (their rate of travel, however, is very slow), decay, or are damped by an anticlockwise wind shift. Pollard found the latter mechanism was necessary to explain the transient nature of the observations, and was hence important in the ocean.

Pollard, Rhines and Thompson (1973) invoked inertial oscillations to explain part of the mixing in the surface layers of the ocean. In a one-dimensional model they assumed the entire mixed layer moved in inertial oscillations under the action of a time-varying wind stress. Since the oscillations are strongly confined to the surface layer, large vertical shears (variations of current speed and direction over small vertical distances) occur at the bottom of the mixed layer. These shears in turn cause instabilities in the flow, which generate turbulence and mix the fluid from below. The authors were successful in explaining the sudden deepening of the mixed layer on the passage of storms.

Buoyancy

The mixed layer is forced by density fluctuations as well as by mechanical stirring, and evaporation is perhaps the most important buoyancy-driving mechanism. Evaporation acts on both the temperature and the salinity to increase the density of the surface water, giving it a tendency to sink convectively. The opposite process--rainfall-- decreases the surface water salinity but may change the sea surface temperature in either direction.

Conservation Relations

One-dimensional mixed-layer models solve simultaneous conservation equations for momentum, buoyancy, heat and salt. For example, P. Niiler and E. Kraus, in Kraus (1981) use, for the buoyancy b,

$$b = -g(\rho - \rho_r)/\rho_r = g\{\alpha(T-T_r) - \beta(s-s_r)\} \tag{17}$$

where ρ is density, T is temperature, and s salinity; the "r" subscript denotes constant "reference" values of quantities in the mixed layer. α and β are the coefficients which describe the (linearized) effect of temperature and salinity on density. Given suitable boundary conditions for the fluxes at the surface and at the base of the mixed layer, all the equations can be solved for the vertical motion, temperature, salinity, and depth of the mixed layer. At the moment the solutions depend on a great deal of mathematical and physical simplification of the turbulent transport processes in the layer. Such use of formulae relating easily-measured quantities to represent complex physical processes is called "parameterization".

Shear Instabilities

In the presence of a stable vertical density gradient, such as that at the base of the mixed layer, instabilities can be produced in the (otherwise laminar) flow by velocity shear, i.e. vertical gradients in the horizontal velocity. Such instabilities are common in nature; they manifest themselves as wavelike "billows" in clouds at the top of the atmospheric mixed layer, for example. That such "billows" exist in the thermocline has been elegantly demonstrated by Woods (1968).

The shear-induced billows generate turbulence, which then enhances the mixing process in the vicinity of the base of the mixed layer, and allows the "mixing down" of fluid to occur much more efficiently than would happen if the mixing was purely convective. Whether or not the billows form is determined by the local ratio of hydrostatic stability (as measured by Δb, the buoyancy difference across the base of the mixed layer), to the shear instability (as estimated by $(\Delta V)^2/h$, where h is an estimate of the thickness of the sheared layer). The ratio is a Richardson Number

$$Ri = \overline{\Delta bh}/(\Delta V)^2 \qquad\qquad (18)$$

Pollard, Rhines and Thompson (1973) compute $\overline{(\Delta V)^2}$ from their
inertial oscillation solutions, thereby assuming that it is
storm-induced oscillations which create shear at the base of the
mixed layer and produce the observed rapid storm-induced deepening of the oceanic mixed layer. The theory is still considered
valid for the initial deepening; other processes are assumed to
dominate at later times. The Richardson number so defined needs
to be less than about unity, for shear instability to be an
important source of mixing energy.

Entrainment

An excellent discussion of entrainment will be found in
O.M. Phillips' contribution to the Kraus (1977) volume. The
process of migration of the boundary between turbulent and
laminar regions, with the turbulence continually incorporating
more and more of the laminar fluid, or "eroding" it, is called
entrainment. It is obviously central to our understanding of
downward mixing in the ocean, but it is in general a highly complex, nonlinear process. For this reason, Phillips and others
have performed a number of elegant laboratory and field studies
that more or less empirically relate the vertical movement of the
bottom of the mixed layer, or "entrainment velocity" w_e, to
various measurable properties of the fluid and its surroundings,
such as the rms (root-mean-square) turbulent velocity shear
$(u^2)^{1/2}/\ell$ where ℓ is a local scale length, and the buoyancy
$g\Delta\rho/\rho$ (ie the Richardson number $g\Delta\rho\ell/u^2$). The entrainment velocity is related to the rate of increase in depth of the mixed
layer, which is usually a prediction of the mixed-layer theories,
available for comparison with experiments.

Advection

Large-scale ocean currents (and spatial inhomogeneities in
such currents) move water horizontally through a given region,
and produce convergences and divergences of heat and salt which
are quite independent of local forcing. Such movement of water
and its properties are accounted for in the equations of motion
by the so-called "advective" terms $\vec{V}.\nabla s$, $\vec{V}.\nabla T$ and $\vec{V}.\nabla V$.

These terms introduce nonlinearities into the equations,
and the basic equations are typically intractable analytically.
Advection is, unfortunately, crucial to any attempt to relate
observations to theory. Without careful investigation, it is

very difficult to discover if an observed time rate of change in
a given quantity has a local cause, or is simply the result of
some large-scale gradient in the quantity being carried through
the observation area. In the study of mixed-layer development in
the ocean advection must always be taken into account. In practice it is generally allowed for in the observational program, by
large-scale surveys of the quantities of interest and the currents which advect them.

Langmuir Circulations

Irving Langmuir (1938) analyzed the wind-aligned streaks, or
"windrows", often found on lakes, giving penetrating insights
into the causative mechanisms. Since then such "Langmuir" circulations (Figure 18) have been observed on all bodies of water.
Because they can cause downwelling velocities as high as 1% of
the wind speed, Langmuir circulations represent a potentially
important mechanism for the downwards mixing of surface waters,
and so have been extensively studied. For a recent review see
the article by R. Pollard in Angel (1976).

The mechanism for their formation (Leibovich and Paolucci,
1980) is now thought to be that described in Figure 19. A small,
initially crosswind perturbation on the downwind surface current
causes crosswind velocity shears, and hence water vorticity as
shown in (a). Coupled with the vertical shear in the water
caused by the Stokes' drift in the wave field (for a description
of the Stokes' drift see the section on Sea Surface Waves), shown
in (b), a vorticity field like that in (c) is set up. This leads
to convergence of surface water (long downwind-oriented lines
where the surface currents are larger than normal), and divergences between. (In the figure, the convergences would coincide
with the maximum of the forward (downwind) moving section of the
initial perturbation (i.e., between 2 & 3 in a) and the divergences, with the upwind-moving perturbation.) The surface currents are acted upon by the wind stress, and since low-momentum
water reaching the surface is accelerated downwind by the surface
stress, the downwind velocities are greatest near the convergences and the forcing is amplified (there is a positive feed-
back). A final equilibrium state is presumably reached in which
the wind stress and the wave field/Stokes' drift it produces and
amplifies are balanced by friction, including turbulent dissipation. The Craik-Leibovich (1976) theory predicts most of the
well-known features of these circulations. It predicts a transfer of energy from small scales (the vertical distances over
which the Stokes' drift shears occur) to large scales, with a
transport efficiency that increases strongly with scale size.
Eventually the Coriolis force must come into play, at which point
Ekman (1905) dynamics must become important, too.

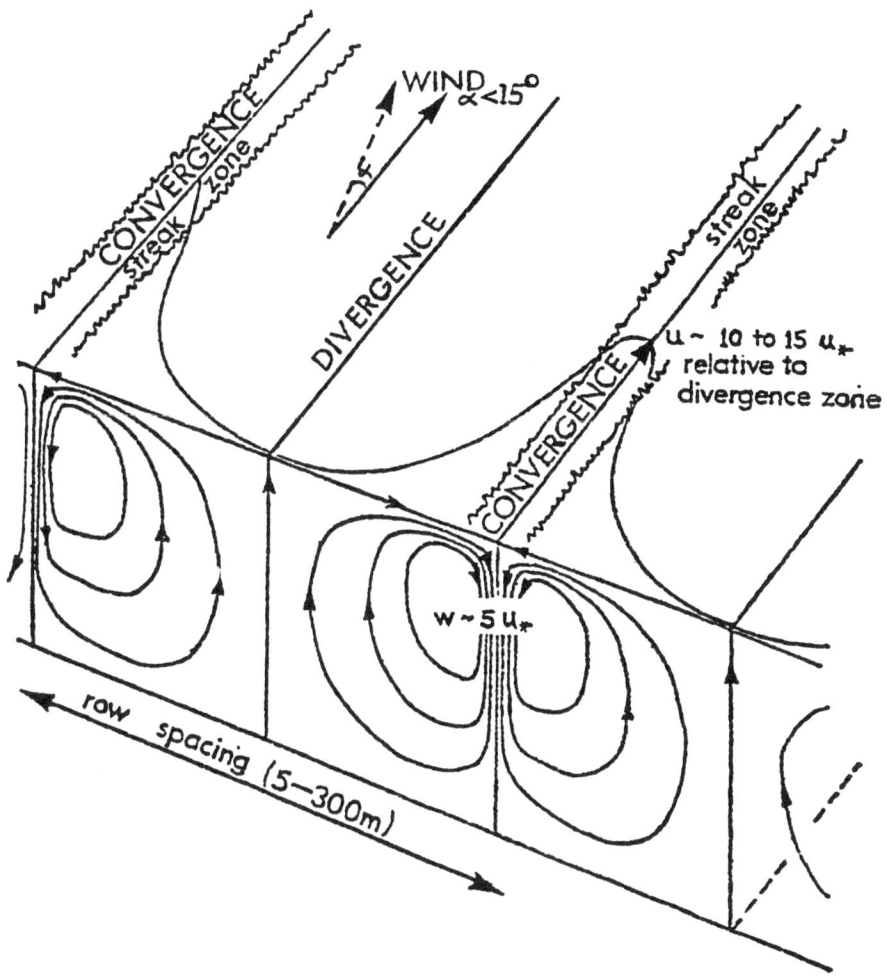

Figure 18. Schematic diagram of the observed structure of
Langmuir circulations. Here $u*_w$ = (wind stress/water
density)$^{1/2}$ is the friction velocity in the water. (After
Pollard in Angel, 1976.)

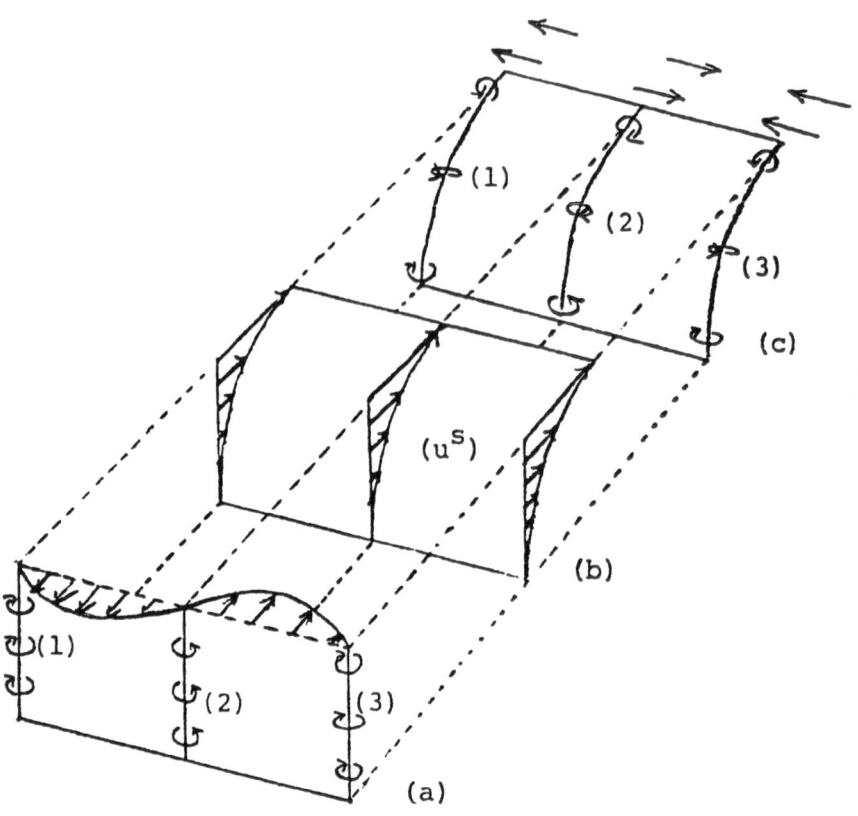

Figure 19. Schematic diagram of the distribution of vorticity in
a wave field caused by a perturbation in surface current. (a) is
the current perturbation; (b) is the Stokes drift u^S associated
with the waves; (c) is the resulting deformation of the vorticity
field, showing how Langmuir "cells" can be set up. The mechanism
is that proposed by Craik and Leibovich (1976); the figure is
from Smith (1980).

Langmuir circulations act as an efficient heat pump, stirring relatively warm surface waters downwards and cool deeper waters towards the surface, where they may gain heat from the air and the sun. Leibovich and Paolucci (1980) define a "mixing efficiency" m_ℓ as the time rate of increase of potential energy of the mixed layer from work done by the Langmuir circulations, divided by the total rate of input of kinetic energy averaged over one Langmuir wavelength in the crosswind and downwind directions (that is, one circuit of a water parcel in the Langmuir "cell"):

$$m_\ell = (d(pe)/dt)/\rho_a \; u_*^3 \; L_c L_d$$

where the "c" and "d" subscripts on the length scales L mean crosswind and downwind. The numerator is in turn given by

$$d(pe)/dt = \int \rho_w \; g\beta_T \; \overline{\hat{w}\hat{\theta}} \; dV$$

where the integral is over the volume dV of the "cell", β_T is the thermal expansion coefficient of water and $\hat{w}\hat{\theta}$ is the rate of vertical tranpsort of heat per unit area caused by the Langmuir circulations. They find values of m_ℓ of 5 to 15 in typical oceanographic situations, as compared to values near 1 for most mixed-layer models. The reader should note that the Leibovich-Paolucci estimate of m_ℓ implicitly assumes that the surface of the entire ocean is covered with Langmuir cells, which of course it is not. To be truly comparable with the values of m_ℓ from the mixed-layer models, the Leibovich-Paolucci m_ℓ estimates should be multiplied by the probability of finding Langmuir circulations at a given place at a given time.

There has not as yet been published an attempt to merge the Leibovich-Paolucci model with a generalized mixed-layer model that includes the effects of solar heating, inertial current mixing, and the larger-scale Ekman circulations. Some such model will be necessary for serious modeling of the dynamics and thermodynamics of the world oceans such as, for example, numerical weather or climate forecasting.

Role of the Mixed Layer in Ocean Dynamics and Climate

The oceanic mixed layer responds to forcing in a manner which has far-reaching dynamic effects. It absorbs three-quarters of the solar radiation reaching the bottom of the atmosphere, and stores it for release at a later time and another

place. The latitudinal variation of rate of storage of heat in
the ocean, according to Oort and Vonder Haar (1976), is compared
in Figure 20 with other terms in the global heat budget. Almost
all the oceanic heat storage occurs above the seasonal thermocline, and so heat storage in the mixed layer can be seen to be a
crucial part of the earth's climate system. On the relatively
sharp density change at the bottom of the mixed layer internal
waves can form and propagate; such waves are found everywhere in
the ocean and serve to make observations difficult. They provide
a widespread mechanism by which energy and momentum can be fed
from the relatively active mixed layer to the less active water
beneath. The mixed layer effectively decouples the deep ocean
from all but the largest-scale, longest-term forcing by the
atmosphere. A good example of this is the well-confined behavior
of storm-forced inertial oscillations (Pollard, 1970). The only
places where this isolation breaks down is at high latitudes
(where strong forced convection sometimes mixes the ocean to
great depths), in upwelling regions near coastlines, and in major
frontal zones, such as at the Antarctic Convergence Zone (see the
Mesoscale Oceanography section).

SEA SURFACE WAVES
Introduction

        Surface waves are perhaps the most obvious and ubiquitous
feature of the oceans. They interact with the air above them and
with the near-surface layer of the ocean beneath, and with each
other. The dynamics of all the interactions have not been worked
out, in spite of more than a century of intensive study.

        Much can be learned about ocean waves by thoughtful inspec-
tion. In the open sea in a strong wind (Beaufort Force 6, for
example), one can find wavelengths from less than a centimeter
(ripples, or capillaries, which make "catspaws" visible on the
sea surface as gusts pass) to hundreds of meters (the largest
waves at Beaufort Force 6 would be about 4 m high from peak to
trough and have wavelengths of about 150 m). The longest wavelengths travel at the greatest speeds and have the largest
heights: the great sea waves can travel at 30 m s^{-1} (60 knots)
and be as high as 30 m. In an actively growing sea the waves
with wavelengths intermediate between the capillaries and the
dominant waves are normally the steepest, and show the greatest
tendency to break. "Breaking" takes many forms, from the turbulent overturning of the crest of a large, long-wavelength wave to
the production of trains of capillary waves moving down the lee-
ward face of steep, short-wavelength waves. The shape of the
larger waves is not quite symmetrical: the wave crests are
sharper than the troughs.

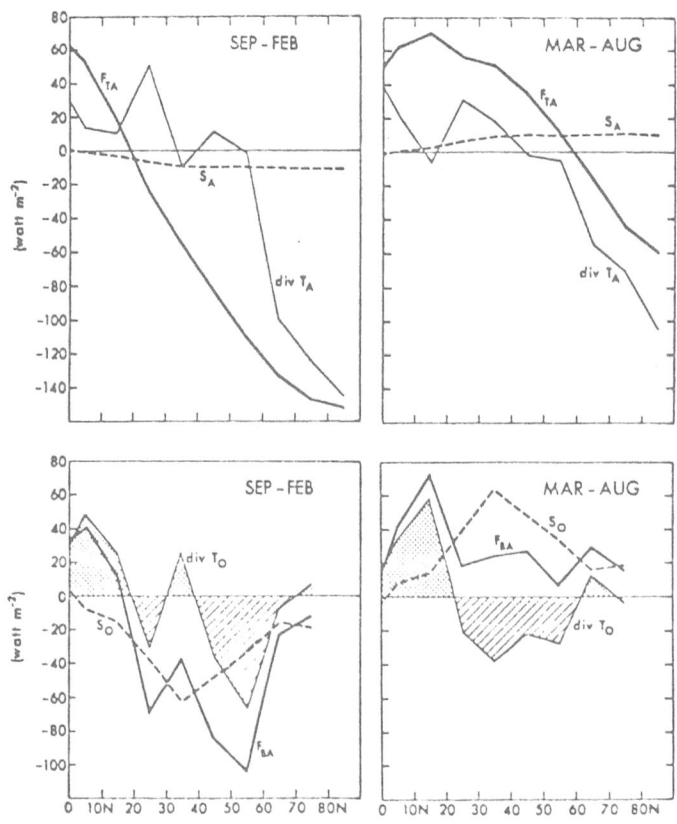

Figure 20. Latitudinal variation of various components of the
earth's heat transport budget for the atmosphere (top) and the
ocean (bottom). The subscript code is T=top, B=bottom;
A=atmosphere, O=ocean. S is "rate of heat storage", F is "vertical flux", and divT is "divergence of transport". Thus S_O
means "rate of storage of heat in the ocean". (After Oort and
Vonder Haar, 1976.)

The dominant waves have a distinct tendency to travel in "groups"; that is, the waves which pass a given point vary their height, three or four large ones normally following after a period of relative calm. If observed carefully, individual waves can be seen "passing through" the groups, indicating that the speed of passage of a given surface undulation is faster than that of the groups. [It turns out that the energy contained in the waves travels at the speed of the wave groups: see the Appendix, relation (A6)]. Also to be seen "passing through" the waves generated by the local wind (the "sea") are trains of relatively long-crested waves not generated locally (known as "swell"). Such waves, after their initial generation by a storm, propagate enormous distances with little change in form.

Storm seas are among the most spectacular of natural phenomena: gigantic mountains of water, travelling at high speed, occasionally breaking at their crests, producing enormous quantities of spray and spume, filling the air with wind-blown spray and the water with spume. To fully appreciate their effect on air-sea exchange of gases and particles, one must experience them.

Waves result when any density discontinuity (e.g., the discontinuity of 1000:1 which is the sea surface) is perturbed. Two forces act to return the sea surface to static equilibrium: gravity, and surface tension. Although they operate in different ways (gravity is a vertical attractive "body" force, while surface tension is an elastic force acting to minimize the surface area of the water), both forces can be included in the equations of motion or boundary conditions of the problem, and mathematical solutions can be found at least for simple cases.

The principal source of difficulty is in specifying the surface boundary conditions. Since the surface is free to move, one must know the solution to the problem (i.e. the shape of the surface) before the boundary condition can be set. A further difficulty arises because the boundary conditions make the equations nonlinear. We will give only some "classical" solutions to linearized problems below and in the Appendix; nonlinear cases are being extensively studied, but are well beyond the scope of this book. In any case, the solutions given will provide an adequate background for understanding most wave phenomena.

Ocean waves occur over a large range of scales, or wave lengths (Figure 21). Their wavelengths vary from millimeters for capillaries to thousands of kilometers for tsunamis (so-called "tidal waves") and the lunar/solar tides (which are forced shallow water waves with lengths of half the circumference of the earth). Ocean waves occur with periods of 0.1 second to 1 day. In this discussion, however, we will limit out treatment to waves

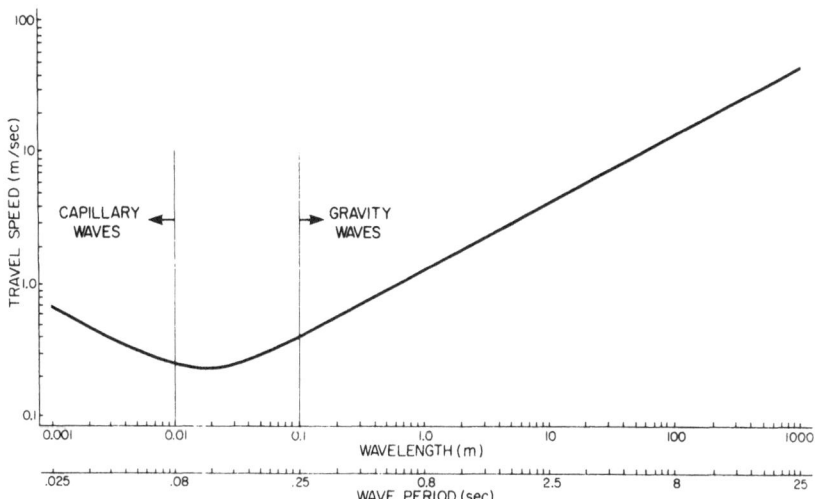

Figure 21. Dispersion relation, or relation between wavelength
or period and travel speed (phase speed) for sea surface waves.
The wave components enclosed by the thin vertical lines are
referred to as gravity/capillary waves. Note that the travel
speed of gravity waves increases with wavelength, while for
capillary waves the travel speed decreases with wavelength.
(After Dobson, 1974.)

generated directly by the wind, i.e. and upper wavelength limit
of about 0.5 kilometers and a longest period to about 20 seconds.
A discussion on tides can be found in the section on Mesoscale
Oceanography.

The amplitude of ocean waves is a strong function of their
wavelength, because water waves become unstable and break if
their steepness (that is, the ratio of wave height to length)
becomes too large. Therefore the longest waves are the largest.
Short wavelength waves are more easily seen than measured (their
visibility is a function of their steepness); typical heights for
10 cm wavelength waves are fractions of a millimeter. The influence of sea waves on the above- and below- water environment
extends only a fraction of a wavelength (wave-induced motions
typically vary as exp $(-k|z|)$, where k is $(2\pi \div$ wave-length) and
$|z|$ is the distance from the interface). This means that in the
air above the sea, where there is usually a mean wind many times
stronger than the wave-induced flows, the wave-induced motions
are hard to detect. In the ocean, on the other hand, where the
mean currents are typically very small, the wave-induced flows
dominate the surface layers.

Under a deep-water gravity wave of amplitude a, individual
parcels of water move in circles with radii a exp kd_o, where
d_o is the mean depth of the particle. Thus a fixed current
meter, located below the troughs, would record a zero mean current in the absence of outside influences: the water in contact
with it would move to an fro with an amplitude proportional to
exp kd_o. On the other hand, a neutrally-buoyant float which
maintained the same mean depth as the current meter, would drift
slowly in the direction of the waves. The effect is caused by
the reduction in the amplitude of wave orbital motions with
depth. Under the crests, at the top of the float's excursion,
its speed in the wave travel direction would be slightly greater
than its return speed under the wave troughs. The phenomenon is
known as the "Stokes drift", after its discoverer. For deep
water gravity waves the Stokes drift speed is given by

$$U_{Stokes} = \omega k a^2 \exp 2kd_o$$

Analysis of the wave equations becomes more complex (and the
solutions closer to reality) for water with nonzero viscosity and
nonzero inherent spin, or vorticity (note that water parcels can
move in circular orbits, as they do beneath waves, and still have
zero vorticity). The effect of viscosity is only evident in very
thin layers near the surface and bottom boundaries, but there,
considerable vorticity can be generated. The thickness δ of
these viscous boundary layers is, approximately,

$$\delta \simeq (2\nu/\omega)^{1/2} \qquad\qquad\qquad\qquad (19)$$

where ν is the so-called "kinematic viscosity" and is equal to
about 10^{-6} m^2 s^{-1} for pure water at 20°C.

In the surface layer, the motion induced by the presence of
viscosity is (naturally enough) a strong function of the surface
boundary condition, that is of the type of stresses applied and
of the surface tension. With no surface stress the (rotational)
perturbation velocities tangential to the wavy surface (u') and
perpendicular to it (w') have amplitudes, in deep water in a
coordinate system moving with the waves (see Phillips, 1977,
pp. 48-49):

$$u' = \delta ak\omega \exp (x'/\delta) \qquad\qquad \text{(in deep water,} \quad (20)$$
$$w' = -2\nu ak^2 \qquad\qquad\qquad\qquad \text{clean surface)}$$

where the prime is for the moving coordinate system. In the
presence of a slick that has essentially no tangential compressibility, the perturbation amplitudes are

$$u' = -\omega a \exp (x'/\delta) \tag{21}$$
$$w' = 2\omega a/k\delta .$$

The attenuation coefficient $\beta = -\partial E/\partial t/2E$, where E is the wave energy density, also varies with the condition of the surface. For a clean surface in deep water (Phillips, 1977)

$$\beta_{\nu} = 2\nu k^2 \qquad\qquad \text{(deep water)} \tag{22}$$

In the presence of a tangentially incompressible film

$$\beta_f = \nu k/2\delta \qquad\qquad \text{(deep water)} \tag{23}$$

When waves are attenuated by surface slicks, account must be taken of the residual wave momentum, since in general, momentum must be conserved. The excess momentum is in fact taken up by an excess velocity (or "streaming") in the surface boundary layer of thickness δ where viscosity is important in the dynamics. In the presence of a tangentially incompressible slick, for instance, the velocity difference ΔU which exists across the boundary layer is (Phillips, 1977)

$$\Delta U = 3/4 \ \omega ka^2 \tag{24}$$

Wave Breaking

As waves of a given wavelength grow by the action of the wind, they eventually reach height-to-wavelength ratios (i.e. slopes) where they become unstable (see e.g. the article by M.S. Longuet-Higgins in Favre and Hasselmann, 1978). With the addition of more energy, they break in a variety of ways. The larger gravity waves tend to spill over at their crests, creating turbulent water in a wake which may extend over a number of wavelengths. A plot of the observed fractional ocean coverage with white caps vs wind speed (Figure 22) indicates the large variability involved. Wave breaking injects bubbles into the water down to a depth of about one wave height (M. Donelan in Favre and Hasselmann, 1978), and as the bubbles burst, they inject spray droplets to heights of one or two wave amplitudes into the turbulent air flow above. The smaller, gravity-capillary waves break in a very different way. They create few bubbles, but the longer-wavelength components among them create small areas of turbulence at their crests; their most characteristic feature is a group of forced capillaries that runs down the front face of the wave, with the shortest-wavelength components leading the longest-wavelength ones.

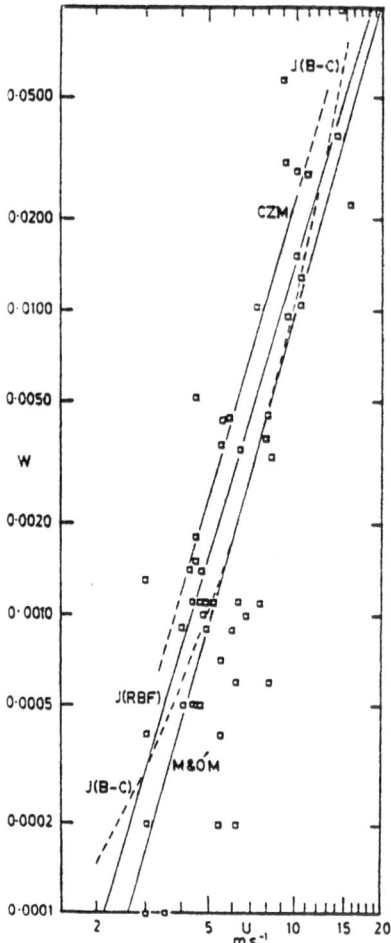

Figure 22. The fraction w of the ocean surface area covered with whitecaps as a function of wind speed u, from observations made during JASIN 1978 (Pollard, 1978) by Monahan and O'Muircheartaigh (JASIN Newsletter No. 25, March 1982). The open squares are the observations; the lines are various earlier empirical curves. The most recent data are fitted best by the central solid line.

Breaking of the short waves is ubiquitous in the real ocean, since short-wavelength gravity waves and long-wavelength capillary waves travel slowly and hence gain energy rapidly from the wind. Very soon after the onset of the wind they reach amplitudes beyond which they cannot grow without breaking (such waves are termed to have reached their "saturation" phase). The phenomenon is easily observed. As the wind begins to steepen the shorter waves, their steepness makes them more visible: the water surface darkens as less and less of its surface becomes available to reflect light from the sun. Patches of short waves formed by wind gusts are referred to as "catspaws", because of their shape. Within a typical catspaw, every wavelet with less than a 10 cm wavelength will be breaking, continuously.

Breaking of the sort described above has been studied theoretically and experimentally by Banner and Phillips (1974). They note that the breaking of small-wavelength waves is strongly influenced by the presence of a shallow, surface-wind-induced layer of highly sheared mean water speed. This layer is caused by the direct frictional influence of the air on the viscous surface layer in the water, and is typically of thickness

$$\delta \simeq 0.5 \ \rho_w \nu / \rho_a u_* \tag{25}$$

where the subscripts "a" and "w" refer to air and water and where u_* (the so-called "friction velocity" in the air: see the section in Interface Dynamics), is defined by

$$u_* = (\tau / \rho_a)^{1/2} \tag{26}$$

where τ is the drag stress per unit area exerted by the air on the water. At sea δ is about 2 mm for a 6 ms^{-1} wind speed. The magnitude of the frictionally induced surface water motion varies along the wave profile; it can be enhanced at the wave crest to many times that in the trough, with the largest enhancements occurring for the slower-moving waves. The point of incipient breaking is that for which the downwind water speed at the wave crest exceeds the phase speed (i.e. the speed at which the disturbance moves through the water). Stokes (1880) calculated the maximum crest elevation before a wave breaks to be

$$\zeta_{max} = C^2 / 2g \tag{27}$$

while Banner and Phillips (1974) found it to be

$$\zeta_{max} = (C - q_0)^2 / 2g \tag{28}$$

where q_o is the downwind drift rate of the surface water at the
point of the wave where $\zeta = 0$.

Wave Generation and Growth

The wave generation process begins, visually, with the
appearance of short-wavelength ripples on calm water, either as
one proceeds away from the beach in an offshore wind or in the
open sea immediately after a wind has sprung up. The initial
waves are composed of a wide spectrum of wavelengths and directions, generated by random fluctuations of air pressure
(Phillips, 1957). Of these initial waves, only those that have a
component of their velocity in the direction of the wind continue
to grow, and their energy (proportional to the square of their
amplitude, a) grows linearly with time, t:

$$a^2(t) \propto \Pi(\omega)\ t\ \cos\theta / \rho_w C$$

where θ is the angle of the waves make with the wind, Π is the
spectrum of turbulent air pressure fluctuations, ρ is water
density, and C is the phase speed of the waves.

Shortly after the wind has begun, and in parallel with the
mechanism just described, a second mechanism becomes active,
which produces short-wavelength waves of a more regular (long-
crested) form. They grow exponentially, because they result from
an instability in the strongly sheared viscous flows within 1-2
mm of the surface in both air and water. Such flows have been
extensively investigated; the most elegant and most convincing
demonstration of their existence and importance in the wave
generation process has been given by the late S. Kawai (1979).
He found that the initial wavelets grew very fast for about 100
wave periods, and then became random in both frequency and direction, losing their long-crested character. The frequency of the
initial wavelets generated was 10 Hz for $u* = 0.1$ ms^{-1} in the
air, and 25 Hz for $u* = 0.3$ ms^{-1}. The process whereby the
waves become random was not elucidated by Kawai. Higher har-
monics of the initial frequencies are clearly visible in the
observed spectra, and in all likelihood the wavelets reach limiting slopes and "break", radiating their surplus energy and
momentum to other wavelengths and directions.

One of the most important aspects of Kawai's work is his
detailed study of the form of the shear flow in the water, which
turned out in his case not to be as predicted by standard viscous
sublayer "theory": the shear was "neither steady nor logarithmic", but rather the water friction velocity varied, starting
from zero and reaching equilibrium with the air friction velocity
(i.e. $\rho_a\ u^2*_a = \rho_w\ u^2*_w$) just prior to the initial

appearance of the waves. An empirical relation was used to
describe the velocity profile in the water, and with it the
author was able to make excellent predictions of the generation
and growth of the wavelets.

Once the waves have been initiated by one of the two mechanisms described, then other effects take over. At high frequencies, where the (capillary/gravity) waves travel slowly relative
to the wind, growth occurs rapidly. In fact, waves of a given
frequency grow so rapidly that they "overshoot" their equilibrium
energy (Figure 23), i.e. they temporarily have slopes larger than
those they can support when in their final equilibrium, or
"saturated" state. In that state, the wave spectrum has the form
(Phillips, 1958)

$$\Phi(f) = \alpha \; g^2 \, f^{-5}$$

where the "constant" α is in fact a weak function of the dimensionless fetch $\tilde{x} = xg/u^2_*$, varying from 5×10^{-2} for $\tilde{x} = 10^5$ to
5×10^{-3} for $\tilde{x} = 10^7$ (Hasselmann et al., 1973).

While the waves are in the "overshot" condition they are in
a nearly-breaking state, and radiate energy and momentum to other
wave spectrum components, both in frequency and directions.
These so-called "nonlinear interactions", investigated in detail
by Hasselmann (1967), are a major source of wave dissipation
(loss of energy to the high-frequency part of the wave spectrum,
and of momentum to the mean current) and growth of the large,
faster-moving components. In fact, the lowest-frequency components in any wind-driven sea are generated in this way, their
growth rate being determined by the shape of the wave spectrum at
frequencies from 1.2 to 1.5 times the frequency at the spectral
peak.

The only effective source of new energy for the gravity
waves in a growing sea is wave-coupled fluctuations in air pressure. The air flow over a wavy surface (Figure 24) speeds up
over crests, causing low pressure (the "Bernoulli effect"), and
slows down over troughs, causing high pressure. In the absence
of the logarithmically-sheared wind profile which is commonly
observed at sea, the Bernoulli pressures generated by the waves
would be exactly out of phase with the wave elevation, and hence
would be zero when the water's vertical velocity is a maximum
(upwards ahead of the crests and downwards ahead of the
troughs). It is pressures in phase with the wave vertical
velocity which are necessary for growth (that is, low pressure
over the upwards-moving water on the front face of the wave, and
high pressure on the downwards-moving rear face). Miles (1957)
has shown that in the presence of a logarithmically-sheared air

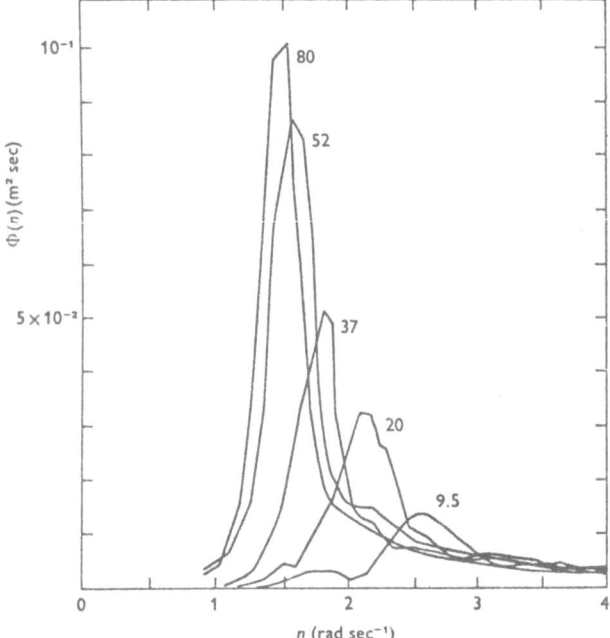

Figure 23. The variation of the wave spectrum with offshore
distance, or fetch. Note the large "overshoot" of waves with
frequencies at or above the local peak frequency, which dis-
appears at the same frequencies at longer fetches as the waves
approach their "saturation" energy. (After Hasselmann et al.,
1973.)

flow typical of that observed over the sea (e.g. Figure 25) the
wave-induced flow is unstable and leads to wave growth.

     To understand how wave-coupled pressures in phase with the
wave vertical velocity are produced, we must view the waves from
a reference frame moving with the wave phase speed. The instability acts as follows (Figure 24). Air particles above the
height where wave speed equals wind speed (the "critical level")
move forward relative to the wave, and those beneath move back-
wards. Air parcels near the critical level move very slowly
relative to the wave, and hence come under the influence of the
wave-induced Bernoulii pressures, mentioned above, which are low
in the region of the crests and high near the troughs. Downwind
from the crests, the adverse pressure gradient in the troughs
causes forward-moving air just above the critical level to be
turned back and downwards, and backward-moving air below to be
turned upwards, forming eddies centered at the crests and
critical level. Because the air flow above the waves is

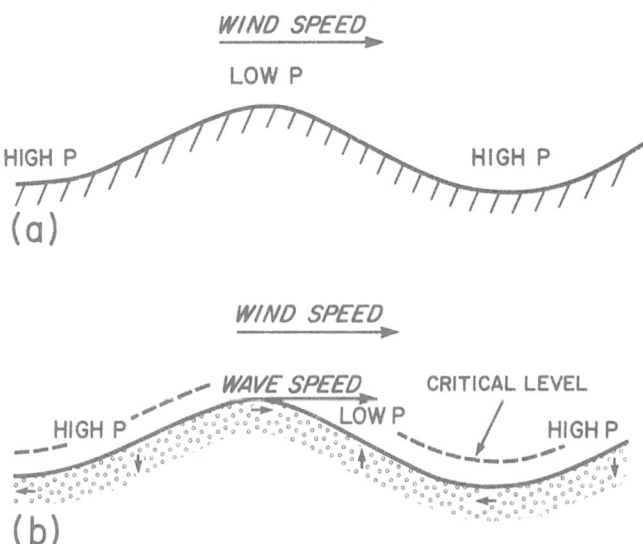

Figure 24. Mean wave-induced air flows and accompanying pressure
fields in the presence of a surface wave, (a) wave speed zero
(i.e. a fixed undulation in a wind tunnel floor; (b) a water wave
propagating downwind with a speed less than that of the wind at a
height of many (>10) wavelengths. The arrows in the water give
the direction of orbital motion of the water as the wave passes
by. Note that the propagation of the wave, by inducing secondary
flows in the air (see the text), causes the pressure field to be
shifted downwind. The pressure field is shown shifted by 1/4 of
a wavelength, at which point the wave growth by pressure forcing
is a maximum.

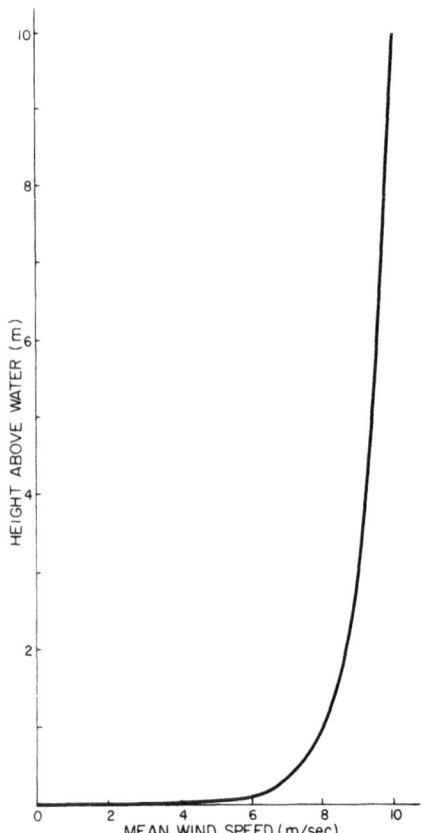

Figure 25. Typical variation of wind speed with height over the ocean. The "mean wind speed" was 10 ms^{-1}, as measured at a height of 10 m. (After Dobson, 1974.)

logarithmically sheared in the vertical, the vorticity, or spin,
of the forward-moving air parcels above the critical height is
slightly less than that of the rearward-moving air below. Therefore downwards-turning air in the troughs ahead of the crests has
a larger radius of curvature than upwards-turning air to the rear
of the crests. As a result the eddies are deformed, moving the
flow streamlines above them forward relative to the waves. The
air flow above encounters wave-coupled pressures which are
shifted forwards, i.e. downwind, causing work to be done by the
air flow on the water.

Mathematically the wave-coupled pressure is expressed as
(Snyder et al., 1981)

$$\hat{p}(t) = \rho_a g \gamma \, \zeta(t)$$

where the hat ($\hat{\,}$) means "wave-coupled" and γ is a dimensionless
complex number

$$\gamma = \mathrm{Re}(\gamma) + i \, \mathrm{Im}(\gamma)$$

with $i = (-1)^{1/2}$; Re and Im stand for "real and imaginary part
of". If the waves grow according to

$$\overline{a^2}(t) = \overline{a_0^2} \, \exp\left\{\mathrm{Im}(\gamma) \, \rho_a \, (\partial E/\partial t)/\rho_w E\right\} t$$

where a_0 is the initial wave amplitude, then Im (γ) is in fact
(ρ_w/ρ_a) times the fractional increase in wave energy per
radian. Snyder et al. (1981) find, empirically,

$$\mathrm{Im}(\gamma) \approx 0.25 \, (\hat{k} \cdot \hat{U}/\omega - 1)$$

where $\hat{k}$ is the wavenumber, and $\hat{U}$ is the mean wind velocity.

Appendix: Definitions

A travelling wave can be defined (Phillips, 1977) in terms
of a sinusoidal disturbance (more complex wave forms are allowed,
providing they can be treated as sums of sinusoids, or Fourier
series) travelling in the direction of their vector wave number
$\hat{k}$, where

$$\hat{k} = 2\pi\hat{l}/\lambda \tag{A1}$$

in which $\hat{l} = \hat{k}/k$ and λ is the crest-to-crest length of the wave.

Such a disturbance may be written as

$$\zeta(\vec{r}, t) = a \cos (\vec{k} \cdot \vec{r} - \omega t) \tag{A2}$$

where ζ is the instantaneous elevation of the sea surface above the mean height at position $\vec{r}$ and time t, a is the wave amplitude (1/2 the crest-trough distance), $\omega = 2\pi f$ is the angular frequency and f is the frequency in Hz, and $\vec{k} \cdot \vec{r}$ means kr $\cos\theta$ where θ is the angle the direction of wave travel makes with the direction chosen for $\vec{r}$. Then, if it is assumed that wave slopes are small (that is, ka $\ll$ 1), it can be shown that a mathematical relation, called the "dispersion relation" exists between the wavelength of a wave and its frequency:

$$\omega^2 (k) = k (g + \gamma k^2 /\rho) \tanh kd \tag{A3}$$

where g is the gravitational acceleration, γ is the surface tension (the force normal to any line on the interface per unit length of the line), ρ is the water density, and d is the water depth.

The result (A3) looks complicated, but it simplifies considerably in particular cases. In "deep water", that is, for $2\pi d > \lambda$, the hyperbolic tangent is very nearly unity (for $d > \lambda/4$, $\tanh kd > 0.99$). Then

$$\omega^2 (k) = k(g + \gamma k^2 /\rho) \qquad \begin{cases} \text{gravity-capillary} \\ \text{waves in deep water} \end{cases} \tag{A4}$$

The disturbances themselves travel at the "phase speed"

$$C = \omega /k = (g + \gamma k^2 /\rho)/\omega \qquad \text{(deep water)} \tag{A5}$$

and transfer their energy at the "group speed" (see the Introduction to the Waves section)

$$C_g = \partial\omega /\partial k = (g + 3\gamma k^2 /\rho)2\omega \qquad \text{(deep water)} \tag{A6}$$

On the right side of (A4) through (A6) the first term is the contribution from gravity and the second from surface tension. A plot of phase speed vs wavelength (Figure 18) shows a minimum in the gravity-capillary region, for which (in clean water) the phase velocity is 0.23 ms^{-1} and the wavelength 0.017 m. One has only to observe how waves "disperse" (that is, separate into their various component wavelengths) to realize that a minimum phase velocity exists. Dropping stones in a pool will show that the capillaries disperse with the short wavelength components running ahead of the longer wavelengths, while the longer gravity waves outrun their shorter companions.

Waves have a mean energy density (i.e., energy per unit area)

$$E = \rho \omega^2 a^2 / 2k \qquad \text{(deep water)} \qquad (A7)$$

which is equally divided between potential and kinetic. For gravity waves ($\gamma k^2 / \rho g \ll 1$)

$$E = \rho g a^2 / 2 = \rho g \, \overline{\zeta^2} \qquad (A8)$$

and for capillary waves ($\gamma k^2 / \rho g \gg 1$)

$$E = \gamma k^2 a^3 / 2. \qquad (A9)$$

Waves also have a momentum density, which is directly related to the energy density by

$$\vec{M} = \vec{\ell} \, \rho \omega a^2 = \vec{\ell} E / c \qquad \text{(deep water)} \qquad (A10)$$

where $\vec{\ell} = \vec{k}/k$ and c is the phase velocity, as before. For small-amplitude pure capillary waves (i.e. $\gamma k^2 / \rho \gg g$) the mean increase in area of surface per unit projected area is,

$$\overline{\Delta A} = \overline{(\partial \zeta / \partial x + \partial \zeta / \partial y + \partial \zeta / \partial z)^2} / 2 \qquad (A11)$$

where terms smaller than $(ka)^2$ have been dropped. For non-small-amplitude waves, $\overline{\Delta A}$ can be large. Cox and Munk (1954) have measured the right side of (A11) in the field; their results are reproduced in Phillips (1977: Fig. 4.23).

GENERAL REFERENCES

Bowman, M.J. and W.E. Esaias (eds.), 1978: Oceanic fronts in coastal processes. Proceedings of a Workshop, SUNY, Stony Brook, N.Y., May 25-27, 1977. Springer-Verlag, 114 pp.

Darwin, G.H., 1962: The tides. W.H. Freeman and Co., San Francisco, 378 pp.

Dobson, F.W., L. Hasse and R. Davis (eds.), 1980: Air-Sea Inter-action: Instruments and Methods. Plenum Press, New York, 801 pp.

Gill, A.E., 1982: Atmosphere-Ocean Dynamics. Academic Press, 662 pp.

Kraus, E.B. (ed.), 1977: Modelling and Prediction of the Upper
    Layers of the Ocean. Pergamon Press, Oxford, 325 pp.

Oceanus, 1976: Ocean Eddies. Woods Hole Oceanographic Institu-
    tion 19, 88 pp.

Oceanus, 1981: Oceanography from Space. Woods Hole Oceano-
    graphic Institution 24, 76 pp.

Phillips, O.M., 1977: The Dynamics of the Upper Ocean (2nd Ed.),
    Cambridge University Press, 326 pp.

Pond, S. and G.L. Pickard, 1983: Introductory Dynamical
    Oceanography (2nd Ed.), Pergamon Press, 329 pp.

Royal Society of London, 1981: Circulation and Fronts in Con-
    tinental Shelf Areas. The Royal Society, London: 177 pp.

Stommel, H., 1960: The Gulf Stream: A Physical and Dynamical
    Description. University of California Press, 202 pp.

Sverdrup, H.U., M.W. Johnson and R.H. Fleming, 1942: The Oceans:
    Their Physics, Chemistry, and General Biology. Prentice-
    Hall, Inc., Englewood Cliffs, N.J. 1087 pp.

Warren, B.A. and C. Wunsch (eds.), 1981: Evolution of Physical
    Oceanography: Scientific Surveys in Honor of Henry Stommel.
    The MIT Press, Cambridge, MA, 623 pp.

REFERENCES

Angel, M., ed., 1977: A Voyage of Discovery: George Deacon
    Memorial Volume. Supplement of Deep-Sea Res., Pergamon
    Press, 696 pp.

Augstein, E., 1981: Atmosphaerische und ozeanische grenz-
    schichten in den niederen breiten. Hamb. Geosphys.
    Einzelschr. 53, 148 pp.

Banner, M.L. and O.M. Phillips, 1974: On the incipient breaking
    of small scale waves. J. Fluid Mech. 65, 647-656.

Batchelor, G.K., 1967: An Introduction to Fluid Mechanics.
    Cambridge Univ. Press, 615 pp.

Bowman, M.J. and W.E. Esaias, 1978: Oceanic fronts in coastal
    processes. Proc. Workshop at Marine Sciences Center, SUNY,
    Stonybrook, NY, May 25-27, 1977. Springer-Verlag, 114 pp.

Broecker, W.S., T. Takahashi, H.J. Simpson, and T.-H. Peng, 1979:
    Fate of fossil fed carbon dioxide and the global carbon
    budget. Science, 206, 409-418.

Budyko, M.I., 1974: Climate and Life. Academic Press, 508 pp.

Cartwright, D.E., 1977: Ocean Tides. In Reports on Progress in
    Physics, 40, 665-708.

Clarke, R.A., and J.-C. Gascard, 1983a: The formation of
    Labrador Sea water. Part I: Large-Scale Processes: J.
    Phys. Oceanogr, 13, 1764-1778.

_____, 1983b: The formation of Labrador Sea water. Part II:
    Mesoscale and Smaller-Scale Processes. J. Phys. Oceanogr.
    13, 1779-1797.

Cox, C.S. and W.H. Munk, 1954: Statistics of the sea surface
    derived from sun glitter. J. Mar. Res. 13, 198-227.

Craik, A.D.D. and S. Leibovich, 1976: A rational model for
    Langmuir circulations. J. Fluid Mech. 73, 401-426.

Csanady, G.T., 1979: The birth and death of a warm core ring.
    J. Geophys. Res., 84, 777-785.

Darwin, G.H., 1962: The Tides. Reissued by W.H. Freeman and
    Co., 378 pp.

Defant, A., 1958: Ebb and Flow: The Tides of Earth, Air and
    Water. U. Michigan Press, Ann Arbor, 121 pp.

Dobson, F.W., 1971: Measurements of atmospheric pressure on
    wind-generated sea waves. J. Fluid Mech., 48, 91-127.

Dobson, F.W., 1974: The wind blows, the waves come. Oceanus,
    17, 29-36.

Dobson, F.W., L. Hasse, and R. Davis (eds.), 1980: Air-Sea
    Interaction: Instruments and Methods. Plenum Press, 801
    pp.

Ekman, V.W., 1905: On the influence of the earth's rotation on
    ocean currents. Ark. Math. Astr. Fys. (Stockholm), 2,
    1-52.

Favre, A. and K. Hasselmann (eds.), 1978: Turbulent Fluxes
    through the Sea Surface, Wave Dynamics, and Prediction.
    NATO Conference Series V, Plenum Press, 677 pp.

Fu, L.-L., and B. Holt, 1982: Seasat views oceans and sea ice
    with synthetic-aperture radar. NASA, Jet Propulsion Lab.
    Pub. 81-120, 200 pp.

Garrett, C.J.R., 1979: Mixing in the ocean interior. Dyn. Atm.
    and Oceans 3, 239-265.

Gill, A.E., 1982: Atmosphere-Ocean Dynamics. Academic Press,
    662 pp.

Gordon, A.L., 1982: World ocean water masses and the saltiness
    of the Atlantic. Presented to Study Conference on
    Large-Scale Oceanographic Experiments in the WCRP, Tokyo,
    May 1982. WMO Secretariat, Geneva.

Gorshkov, S.G. (ed.), 1978: World Ocean Atlas, Vol. 2, Atlantic
    and Indian Oceans. Pergamon Press, Oxford.

Hasselmann, K., 1967: Nonlinear interactions treated by the
    methods of theoretical physics (with application to the
    generation of waves by wind). Proc. Roy. Soc. A, 299,
    77-100.

Hasselmann, K. et XV al., 1973: Measurements of wind wave growth
    during the Joint North Sea Wave Project (JONSWAP).
    Deutsches Hydrogr. Zeitschr. A., 12, 95 pp.

Holland, W.R., 1978: The role of mesoscale eddies in the general
    circulation of the ocean-numerical experiments using a wind-
    driven quasi-geostrophic model. J. Phys. Oceanogr. 8, 363-
    392.

Holland, W.R., and P.B. Rhines, 1980: An example of eddy-induced
    ocean circulation. J. Phys. Oceanogr. 10, 1010-1031.

Jerlov, N.G., 1976: Marine Optics. Elsevier Press, 231 pp.

Kawai, S., 1979: Generation of initial wavelets by instability
    of a coupled shear flow and their evolution to wind waves.
    J. Fluid Mech. 93, 661-703.

Knudsen, M., 1901: Hydrographical Tables. G.E.C. Gad,
    Copenhagen. Tables for the calculation of sigma-t from
    values of salinity and temperature, p. 63.

Kraus, E.B. (ed.), 1977: Modelling and Prediction of the Upper
    Layers of the Ocean. Pergamon Press, 325 pp.

Langmuir, I., 1938: Surface motion of water induced by wind.
    Science 84, 119-123.

Leibovich, S. and S. Paolucci, 1980: The Langmuir circulation instability as a mixing mechanism in the upper ocean. J. Phys. Oceanogr. 10, 186-207.

LeBlond, P.H. and L.A. Mysak, 1978: Waves in the Ocean. Elsevier, 602 pp.

Miles, J.W., 1957: On the generation of surface waves by shear flows. J. Fluid Mech. 3, 185-204.

Munk, W.H., 1950: On the wind-driven ocean circulation. J. Meteorol. 7, 79-93.

Munk, W.H., 1966: Abyssal recipes. Deep-Sea Res. 13, 707-730.

Oceanus, 1976: Ocean Eddies. Woods Hole Oceanographic Institution 19, 88 pp.

Oort, A.B. and T.H. Vonder Haar, 1976: On the observed annual cycle in the ocean-atmosphere heat balance over the Northern Hemisphere. J. Phys. Oceanogr. 6, 781-800.

Parker, C.E., 1971: Gulf Stream rings in the Sargasso Sea. Deep-Sea Res. 18, 981-993.

Phillips, O.M., 1957: On the generation of waves by turbulent wind J. Fluid Mech. 2, 417-445.

Phillips, O.M., 1958: The equilibrium range in the spectrum of wind-generated waves. J. Fluid Mech. 4, 426-434.

Phillips, O.M., 1977: The Dynamics of the Upper Ocean. Cambridge Univ. Press, Second Ed. 336 pp.

Pollard, R.T., 1970: On the generation by winds of inertial waves in the ocean. Deep-Sea Res. 17, 795-812.

Pollard, 1978: The Joint Air-Sea Interaction Experiment - JASIN 1978. Bull. Amer. Meteorol. Soc. 59, 1310-1318.

Pollard, R.T., P.B. Rhines, and R.O.R.Y. Thompson, 1973: The deepening of the wind mixed layer. Geophys. Fluid Dyn. 3, 381-404.

Pond, S. and G.L. Pickard, 1983: Introductory Dynamical Oceanography (2nd Ed.). Pergamon Press, 329 pp.

Richardson, P.L., R.E. Cheney, and L.A. Martini, 1977: Tracking a Gulf Stream ring with a free drifting surface buoy. J. Phys. Oceanogr. 7, 581-590.

Robinson, M.K., R.A. Bauer, and E.H. Schroeder, 1979: Atlas of
    North Atlantic – Indian Ocean monthly mean temperatures and
    mean salinities of the surface layer. U.S. Naval
    Oceanographic Office Ref. Pub. 18, Washington, D.C.

Roether, W., 1982: Transient Tracers in the Ocean. Contribu-
    tion, Joint CCCO/JSC Study Conference on WCRP Oceanography,
    Tokyo, 1982. WMO WCRP White Paper, WMO Secretariat Geneva.
    To be published.

Rossby, C.-G., 1938: On the mutual adjustment of pressure and
    velocity distributions in certain simple current systems,
    II. J. Mar. Res. 1, 239–263.

Simpson, J.H. and J.R. Hunter, 1974: Fronts in the Irish Sea.
    Nature 250, 404–406.

Smith, J.A., 1980: Waves, currents and Langmuir circulation.
    Ph.D. Dissertation, Inst. Oceanography, Dalhousie University
    242 pp.

Snyder, R.L., F.W. Dobson, J.A. Elliott and R.B. Long, 1981:
    Array measurements of atmospheric pressure above surface
    gravity waves. J. Fluid Mech. 102, 1–59.

Stokes, Sir G.G., 1880: Math. and Physical Papers 1, Cambridge
    University Press, 314–326.

Stommel, H., 1948: The westward intensification of wind-driven
    ocean currents. Trans. Amer. Geophys. Union 29, 202–206.

Stommel, H., 1960: The Gulf Stream, Univ. Calif. Press:
    202 pp.

Stommel, H. and A.B. Aarons, 1960: On the abyssal circulation of
    the world ocean – II. An idealized model of the circulation
    pattern and amplitude in oceanic basins. Deep-Sea Res. 6,
    217–233.

Sverdrup, H.U., 1947: Wind-driven currents in a baroclinic
    ocean; with application to the equatorial current of the
    eastern Pacific. Proc. Nat. Acad. Sci. Wash. 33, 318–326.

Sverdrup, H.U., M.W. Johnson, and R.H. Fleming, 1942: The
    Oceans. Prentice-Hall, Inc., Englewood Cliffs, N.J. 1087
    pp.

Tabata, S. and L. Giovando, L., 1963: The seasonal thermocline
    at Ocean Weather Station "P" during 1956 through 1959.
    Fish. Res. Board Can., MS Rept. Ser. 157, 27 pp.

Turner, J.S., 1979: Buoyancy Effects in Fluids. Cambridge
    Univ. Press, 368 pp.

UNESCO, 1981a: Background Papers and Supporting data on the
    Practical Salinity Scale 1978. UNESCO Technical Papers in
    Marine Science 37, UNESCO, Paris.

UNESCO, 1981b: Background papers and supporting data on the
    International Equation of State of Seawater, 1980. UNESCO
    Technical Papers in Marine Science 38, UNESCO, Paris.

Warren, B.A. and C. Wunsch (eds.), 1981: Evolution of Physical
    Oceanography. MIT Press, 623 pp.

White, W.B. and J.P. McCreary, 1976: On the formation of the
    Kuroshio meander and its relationship to the large-scale
    ocean circulation. Deep-Sea Res. 23, 33-47.

Woods, J.D., 1968: Wave-induced shear instability in the summer
    thermocline. J. Fluid Mech. 32, pp. 791-800.

Woods, J.D., 1980: Diurnal and seasonal variation of convection
    in the wind-mixed layer of the ocean. Quart. J. Roy.
    Meteorol. Soc. 106, 379-394.

Woods, J.D., 1983: Climatology of the upper boundary layer of
    the ocean. In Proceedings of Joint JSC/CCCO Joint Study
    Conf. on Oceanographic Experiments in the WCRP, Tokyo, 10-21
    May 1982. WCRP Publication Series, No. 1, WMO, Geneva, p.
    147 ff.

World Meteorological Organization, 1981: On the assessment of
    the role of CO_2 on climate variations and their impact. WMO
    Secretariat WCRP White paper No. 3, 29 pp.

Worthington, V., 1969: An attempt to measure the volume
    transport of Norwegian Sea overflow water through the
    Denmark Strait. F.C. Fuglister 60th Anniv. Vol. Deep-Sea
    Res. 16 (Supplement), 421-432.

Wunsch, C., 1981: The promise of satellite altimetry. Oceanus
    24, 17-26.

ANNOTATED BIBLIOGRAPHY

Meteorological and Oceanographic Measurements, and Sampling
Strategy

   This annotated bibliography contains references useful for
those interested in experiment design, including available
instruments, techniques of measurement, and the statistics of
sampling, error analysis, and spectral analysis. If it seems
weighted towards the oceanic side, this is because meteorological
instruments and techniques are more standardized. Meteorological
measurements are organized into a global weather analysis network
by the World Meteorological Organization, Geneva. The WMO
publishes many handbooks on meteorological measurements, which
are available from the national weather bureaus.

Baker, D.J., 1981: Ocean instruments and experiment design. In
     Evolution of Physical Oceanography, B.A. Warren and C.
     Wunsch (eds.), MIT Press, 396-433.

This article gives a complete treatment of deep-sea physical
oceanographic instruments developed since publication of The
Oceans (see below), and goes into many of the advantages and pit-
falls associated with their use. It does not discuss remote
sensing, and leaves surface measurements to Dobson, Hasse and
Davis (see below).

Bernstein, R.L. (ed.), 1982: Seasat Special Issue I: Geo-
     physical Evaluation. J. Geophys. Res. 87 (C5), 3175-3438.

This series of papers summarizes the state of the art in micro-
wave remote sensing of the sea surface attained by SEASAT I. It
contains review articles on the radar altimeter, passive micro-
wave radiometer, radar scatterometer, and synthetic aperture
radar. Neither infrared sensing of sea surface temperature no
color scanners are mentioned (they were not part of SEASAT).

Blackman, R.B. and J.W. Tukey, 1959: The Measurement of Power
     Spectra. Dover, Inc., New York, 190 pp.

This slim volume has long been the bible of those interested in
designing optimal sampling strategies and collecting data for
spectral analysis. The style is terse, the notation is curious
and at times hard to follow, but it is all there for the determined reader.

Dobson, F.W., L. Hasse and R. Davis (eds.), 1980: Air-Sea Inter-
     action: Instruments and Methods. Plenum Press, 801 pp.

This book deals in considerable detail with both the instruments
and the measurement techniques in use today in the field of air
sea interaction. Measurements in both air and sea are treated.
There is a chapter on sampling the surface microlayer, as well as
sections on velocity, temperature, salinity, pressure, and
instrument platforms. Each section is written by an acknowledged
expert. Remote sensing is not included.

Gower, J.F.R. (ed.), 1981: Oceanography from Space. Plenum
     Press, Marine Science Series 13, 978 pp.

This book is the proceedings of an international symposium on the
subject, held in Venice in 1980. It is not truly a reference
work, but rather a collection of papers, as is the SEASAT issue
edited by Bernstein. The reader must therefore dig for information on how to use remote sensing as a tool for investigating the
ocean. Sections are included on infrared measurements of sea
surface temperature, ocean color scanners, and sea ice sensing,
as well as on the subjects covered by the Bernstein reference.

Jenkins, G.M. and D.B. Watts, 1969: Spectral Analysis and its
     Applications. Holden-Day, Inc., San Francisco, 525 pp.

This book includes and builds on the information contained in
Blackman and Tukey. It is less of a "handbook" and more of a
text, and delves into some of the underlying theory of sampling
and spectral analysis, as well as probability theory and statistics. The only item important to this audience which is not
covered is the Cooley-Tukey Fast Fourier Transform.

Sverdrup, H.U., M.W. Johnson and R.H. Fleming, 1942: The Oceans:
     Their Physics, Chemistry and General Biology. Prentice-
     Hall, Englewood Cliffs, NJ, 1087 pp.

Long the only general source of information on the oceans, this
book remains the oceanographer's bible. Chapter 10: Observations and Collections at Sea, is an excellent source of information on the "classical" instruments and techniques.

INDEX

123